Social Poison

Social Poison

The Culture and Politics of Opiate Control in
Britain and France, 1821–1926

HOWARD PADWA

The Johns Hopkins University Press
Baltimore

© 2012 The Johns Hopkins University Press
All rights reserved. Published 2012
Printed in the United States of America on acid-free paper
2 4 6 8 9 7 5 3 1

The Johns Hopkins University Press
2715 North Charles Street
Baltimore, Maryland 21218-4363
www.press.jhu.edu

Library of Congress Cataloging-in-Publication Data

Padwa, Howard.
Social poison : the culture and politics of opiate control in Britain and France, 1821–1926 /
Howard Padwa.
  p.; cm.
Includes bibliographical references and index.
ISBN-13: 978-1-4214-0420-2 (hardcover : alk. paper)
ISBN-10: 1-4214-0420-6 (hardcover : alk. paper)
1. Opioid abuse—Great Britain—History—19th century. 2. Opioid abuse—France—
History—19th century. 3. Opioid abuse—Great Britain—History—20th century.
4. Opioid abuse—France—History—20th century. 5. Drug control—Great Britain—History—
19th century. 6. Drug control—France—History—19th century. 7. Drug control—Great Britain—
History—20th century. 8. Drug control—France—History—20th century. I. Title.
[DNLM: 1. Opioid-Related Disorders—history—France. 2. Opioid-Related Disorders—history—
Great Britain. 3. Drug and Narcotic Control—history—France. 4. Drug and Narcotic Control—
history—Great Britain. 5. History, 19th Century—France. 6. History, 19th Century—Great Britain.
7. History, 20th Century—France. 8. History, 20th Century—Great Britain. 9. Public Opinion—
history—France. 10. Public Opinion—history—Great Britain. WM 11 FA1]
HV5840.G7P33 2012
362.29′3094109034—dc23   2011021308

A catalog record for this book is available from the British Library.

*Special discounts are available for bulk purchases of this book. For more information,
please contact Special Sales at 410-516-6936 or specialsales@press.jhu.edu.*

The Johns Hopkins University Press uses environmentally friendly book materials,
including recycled text paper that is composed of at least 30 percent post-consumer waste,
whenever possible.

To my grandmothers, May Baum and Betty Padwa

CONTENTS

It would be impossible to acknowledge all of the individuals and organizations whose support has helped bring this book to fruition, but a few deserve special mention.

First, I thank Richard Romero, the high school teacher who changed the way I viewed the world and its past and inspired me to pursue the study of history. At the University of Delaware, professors Suzanne Austin, Laurence Duggan, W. O. Maloba, and David Shearer provided me with encouragement to continue in my academic career, and at the University of California, Los Angeles, Joel Braslow, Adrian Favell, Lynn Hunt, Marcia Meldrum, and Dora Weiner shared their expertise and provided support as I carried out my research. Peter Baldwin deserves special thanks for motivating and encouraging me throughout the process, while sharing invaluable insights and ideas that helped push the project along.

Generous support from the UCLA Center for European and Eurasian Studies, the UCLA Department of History, the American Institute for the History of Pharmacy, Égide, the Institute for Humane Studies, and the University of California Humanities Research Institute gave me the opportunity to research and write the manuscript. In Paris, the staff at the Bibliothèque médicale Henri Ey at Sainte Anne Hospital, the Académie de médecine, and the Archives Nationales provided assistance as I began my research. Clotilde Carrandié at the Bibliothèque médicale de Marmottan was particularly helpful in steering me toward sources I probably would not have found without her guidance. During my time in Paris, Emmanuelle Retaillaud-Bajac also generously shared her expertise on the history of French drug policy. In Britain, staff at the British Public Record Office at Kew, the British Library, and the Wellcome Library helped guide me as I began my research there. Jacqueline Wehmueller at the Johns Hopkins University Press has been extraordinarily helpful in shepherding me through the editorial process, and my reviewer at the Press provided constructive and detailed critiques that guided me as I revised the book. My copyeditor, Linda Strange, also deserves special thanks for helping me brush up the final manuscript.

Finally, I thank my friends and family, particularly my parents, Stephen and Linda Padwa, for sticking with me through thick and thin and giving me a lift whenever I needed it. Most of all, I owe tremendous gratitude and thanks to my lovely wife, Joy Moini, whose friendship, companionship, and love are better than any high I could possibly imagine.

Social Poison

A Tale of Two Drug Policies

For an otherwise law-abiding morphine addict struggling to overcome addiction in the late 1920s, Britain was a more welcoming place than France. Although narcotics were subject to similar controls in both countries, the British government was willing to permit maintenance treatment in intractable cases of morphine and heroin addiction, but the French government usually was not. British doctors had the freedom to provide nontapering doses of opiates to confirmed addicts, provided they followed proper precautions. The French authorities, however, saw such maintenance as an indulgence of addiction rather than an appropriate way to treat it, and they were not reluctant to take action against doctors and pharmacists who provided opiates to addicts. Any French doctor who wrote a prescription for more than one week's worth of opiates to ease the pains of withdrawal was liable to prosecution. Thus, whereas some Britons had the option to continue leading relatively normal lives in spite of their addictions, their French counterparts were trapped in a state of medical and legal uncertainty, constantly compelled either to undergo tremendous physical suffering if they quit using drugs or to risk breaking the law.

One of the aims of this book is to explain why and how this difference between the two countries arose. British and French policies concerning opiate addiction were created during the period surrounding World War I. But attitudes about opiate use, and the addictive disorders it often caused, had their roots in the Victorian era, when both the allures and the dangers of the drugs captured the attention of writers, doctors, and, eventually, policymakers. The drug discourses that developed over the course of the nineteenth century in Britain and France shaped the thinking about opiates well into the twentieth century, influencing the timing, content, and character of the national drug policies that would emerge in the 1910s and 1920s. Changes in medical knowledge, culture, politics, and drug-using demographics merged to create distinct ways of understanding the problems that opiate use could cause, transforming it from an issue that concerned only the health and character of addicts into one that, if left uncontrolled, could have widespread ramifications for the British and French nations as a whole. I explore this transformative process, comparing the different paths taken by Britain and France in developing their drug control regimes at the beginning of the twentieth century.

The Historical and Comparative Analysis of Drug Policy

The historical literature on narcotics and narcotics control in the first half of the twentieth century has become increasingly rich in recent years, yet there has been little comparative work examining differences in national drug control policies of the period. Jan-Willem Gerritsen's *The Control of Fuddle and Flash*, which discusses early control policies in Britain, the Netherlands, and the United States, is the only explicitly comparative history of nineteenth- and early-twentieth-century drug policies in the industrialized world, though it focuses more on the history of alcohol than other psychoactive substances.[1] Other researchers have integrated comparative points into their studies of substance abuse policies from this era, though generally as an aside, not as a main focus of their research.

Most drug policy comparisons looking at the early twentieth century focus on Britain in the 1920s, probably because more recent critics of punitive drug policies have heralded the so-called British System as an exemplar of restraint, reason, and compassion.[2] Comparative discussions of British addiction policy generally fall into two camps. Some authors frame the divergence between British and other national drug policies as a philosophical one, claiming that whereas Britain developed a more "medical" approach to addiction, other countries were more moralistic or chose to marginalize medical input during their policymaking processes.[3] Other researchers suggest that the differences between British and other national policies were not so much ideological as demographic. They argue that the British were willing to make concessions allowing maintenance treatment primarily because their addicts were less numerous, were of higher socioeconomic standing, and used drugs in a less socially menacing manner. The British, therefore, were able to sanction maintenance treatments, not because they viewed the drug problem differently, but because they had a different sort of drug problem altogether. More detailed research on British drug policy from the 1920s through the 1960s tends to support this line of thinking, as scholars have found that Britain's tolerance of maintenance treatments was not necessarily more "medically" oriented than the policies of other nations and that it allowed looser prescribing practices to exist only "within a penal framework" that still repressed drug trafficking and use.[4]

My aim is not to refute previous historical comparisons of early-twentieth-century drug policies in Britain or elsewhere; rather, it is to refine and enrich these accounts by undertaking a more rigorous comparison, looking at the evolution and formulation of drug policies in two countries that took divergent approaches to addressing the drug problem: Britain and France. The book focuses not on all controlled substances but on those that represented the central front in the early war on drugs: opium, opiates, and semisynthetic opiates (mainly opium, morphine, and heroin).[5] There were several reasons for my decision to focus on these drugs.

First, in the course of my research, I found that opiates were the main targets of much of the national anti-drug legislation that emerged from London and Paris in the first two decades of the twentieth century, and of the international control regime that took shape during the same period. To understand the dynamics behind drug control at this time, it is crucial to understand what it was about opiates that grabbed the attention of policymakers worldwide. Second, compared with other drugs that became subject to tighter controls during this era (e.g., cocaine, cannabis), opiates had a long and elaborate history in Britain and France. Because of the social and cultural meanings assigned to opiates during the nineteenth century, their recreational use came to represent more than a mere indulgence or exotic alternative to alcohol. Opiates developed a distinctive character in nineteenth-century British and French drug discourses, one that would be of critical importance when narcotics control became a matter of policy concern in the twentieth century. Third, opiate addiction, much more than addiction to other tightly controlled substances, was as linked to medical use as it was to recreational indulgence, and it was also considered more treatable. Thus, though it had distinct cultural and social significance, opiate addiction was still considered a medical condition with a potential medical cure. Consequently, opiate addiction posed a unique challenge for policymakers, who needed to strike a balance between the penal system's need to punish individuals who violated drug controls and society's medical and moral mandates to provide compassionate treatment for the victims of disease.

As for my choice of countries, France is an appropriate candidate for comparison mainly because the story of its early drug policies has scarcely been studied, though Emmanuelle Retaillaud-Bajac's *Les paradis perdus* has largely filled this historiographical void. Britain offers a good contrast to France because, in many respects, its history is similar: the countries are neighbors not only geographically but, in the story of drug control, temporally. Both promulgated their landmark regulations limiting the flow of opiates—the 1916 poisonous substances law in France, and Defence of the Realm Act regulation 40B in Britain—during the same month in the summer of 1916. Yet, in spite of surface similarities, the two countries took different approaches to the drug problem, as recognized by historians of French drug policy but not yet explored in detail.[6]

In part, the comparative analysis developed here builds on previous works on the history of early-twentieth-century drug policy in industrialized nations. Some works have shown how concerns about the health of drug users and the harms that addicts could cause others motivated states to check the free flow of narcotics, as did moral and religious disapproval of drug habits.[7] Through processes of medicalization and normalization, medical men were able to translate moral biases against drug use into medical theories of addiction, labeling recreational drug use a "deviant" practice and then seeking to eliminate it.[8] Social and cultural factors only tangentially related to the dangers posed by narcotics themselves also played a role in prompting the marginalization and penalization of drug users. Many historians have shown that as excessive opiate use fell out of favor in the medical community and practitioners began prescribing the drugs less liberally, drug-induced debauchery came to be associated with urban vice, the lower classes, racial minorities, and foreigners, thus rallying opposition to unrestricted use and limiting resistance to laws aimed at controlling and punishing users.[9] Furthermore, the medical, psychiatric, and pharmaceutical professions used addiction to carve out professional niches that could advance their own interests and, in the process, give them a monopoly over the power to dispense narcotics to the general public.[10] Fears that the free flow and unlimited consumption of narcotics could endanger the future of the social collectivity were also prominent, and such concerns undergirded states' drives to institute more stringent controls over psychoactive substances.[11]

Beyond the historiography looking at the early twentieth century, research comparing more recent developments in drug policy also provides some valuable insights into what has led different countries to tackle the drug problem in different ways. Restrictions on controlled substances are similar in most industrialized nations, but the ways in which policies are put into practice differ on questions of how to implement law enforcement, prevention, and treatment measures.[12] Sociologists and policy analysts have shown how institutional factors, as well as the relationship between national and local jurisdictions, influence the development of law enforcement and treatment measures, as do health and welfare policies, traditions regarding civil liberties, and the demographics of drug-using populations.[13]

Differences in political culture and discourse also shape how states respond to their drug problems, creating different ways of interpreting what sort of threat drugs pose to national populations and different justifications and rationalizations for criminal justice and addiction treatment policies.[14] Scholars comparing contemporary drug policies are beginning to more closely explore the connections between drug discourses and divergent enforcement and treatment practices, showing that in addition to more tangible policy traditions and demographic factors, national ways of thinking about drug use and the threat it poses to the national collectivity have shaped the formation and implementation of policy. Tim Boekhout van Solinge, in particular, has studied how distinctly national "drug policy discourses" have been related to policy developments in France, Sweden, and the Netherlands since the 1970s.[15] The French, he found, viewed the drug problem through the lenses of national security and public order, leading them to emphasize morality and citizenship in their drug control system. The Swedish, by contrast, framed drug use as a cultural issue to be dealt with through social policy and measures designed to contain the spread of drug problems, and the Dutch viewed addiction as a public health matter to be handled with policies that reduce the harms associated with drug use. Thus, while differences in the scope, scale, and character of each nation's drug problem can explain some differences in drug policy, cultural, historical, and political contexts also play a role in shaping how states respond to the legal and public health challenges posed by drug use.[16]

Using lessons from transnational analyses of more recent drug policies as a guide, my aim is to adopt a similar approach in comparing drug policies from a century ago, looking at both tangible and intangible differences that could account for the divergences between Britain and France. Some factors proved more important than others in setting British and French drug policies on different trajectories. In many respects, the history of narcotics control in the two countries followed a similar path. Medical researchers in both nations recognized and decried the dangers of addiction at about the same time, and writers on both sides of the Channel conflated moral censure of drug abuse with scientific claims against it. Britain and France controlled opiates with pharmacy laws in the nineteenth century, then more narrowly tailored drug laws in the twentieth, meaning that both countries saw licit control over the distribution of drugs come under medical and pharmaceutical control around the same time. There were, however, two major differences that fell along national lines throughout the late nineteenth and early twentieth centuries—one discursive, the other demographic.

Two Nations, Two Drug Problems

The divergence between British and French drug policies did not stem simply from the British treating addiction as a "medical" issue and the French viewing it as a "penal" one. Rather, the social and political dangers posed by opiate use were discussed differently in the two countries, as their medical researchers, writers, and policymakers had different ways of conceptualizing the questions brought up by the unregulated flow of opiates and its possible effects on their national communities. Broader and distinctly national ways of thinking about society and the types of problems that could threaten social order infiltrated how the British and French framed drug use and its consequences in the late nineteenth and early twentieth centuries. Consequently, a discursive divide emerged between the anxieties about narcotics that were articulated in Britain and those that arose in France throughout the era.

In Britain, where the national community was imagined as one defined by liberty, individualism, and the functioning of the free market, opiates represented two distinct, yet interrelated threats. First, they were believed to make users apathetic and lazy, traits that ran counter to the capitalist ethos that helped define the British way of life. Second, opium's traditional associations with the Chinese made it seem particularly menacing for the British, since in late-nineteenth-century British culture, the Chinese represented the antithesis of the British way of life. Thus opium posed a threat not only to the physical health of Britons but also to their ideological health, by compromising their senses of industriousness, self-sufficiency, and, most importantly, Britishness. In British works of fiction and anti-opium propaganda, opium was depicted as a substance that had the power to transform upstanding Britons into individuals who were indolent, lazy, and tainted by their connection with "the Orient." Thus British conceptions of good citizenship—along with ideas of what a bad citizen would look like—were grafted onto perceptions of opiates, creating a fear of the drugs that transcended universal concerns about public health and morality. Beyond poisoning addicts themselves, opiates came to be seen as potentially poisonous for the national community as a whole and for the essence of British citizenship as it was then understood.

In France, society was imagined in a more collectivistic way, and engagement of citizens with the social whole was considered a fundamental prerequisite of proper sociopolitical organization. Good French citizens were expected to be devoted and civic-minded. There was little tolerance for opiates, which were believed to both represent and engender unbridled individualism and solipsism. French concerns about the effects that opiates could have on the spiritual and ideological health of the nation were expressed in a series of late-nineteenth-century novels and vignettes describing the potentially disastrous effects of opiates on military morale and devotion. These anxieties were reinforced by a 1907 espionage case involving a would-be traitor who blamed opium for his actions. As in Britain, opiates were poisonous to the body not only of he who indulged in them but of the entire nation, whose very existence, in France, was understood to depend on the civic engagement of loyal republican citizens—a devotion that was seen as incompatible with opiate use.

This is not to say that Frenchmen were not worried by the tangible repercussions of addiction or that the British were oblivious to the effects of opiate use on their sense of community. But in general, the relationship between opium and citizenship formed a unifying thread among French concerns about the drugs, and worry over the economic costs of the widespread availability of opiates was a recurrent theme in Britain. National styles of thinking about citizenship thus permeated the more universal fears about health and morality that narcotics struck in the hearts and minds of Britons and Frenchmen. Opium came to be seen, as one politician termed it in 1913, a "social poison" that had the potential to undermine ideological health by turning upstanding members of the national community into bad citizens.[17] Since understandings of citizenship differed in Britain and France, so did conceptions of how and why opiates were inimical to good citizenship. Consequently, the two countries developed their own brands of what I call "anti-narcotic nationalism," reasons for wanting to restrict drug use that were both specifically *national* and *anti-narcotic*, built on assumptions about what bound the national community together and on a fear of what opiates could do to society if their habitual use proliferated.

Anti-narcotic nationalism transcended the realm of discourse to influence policy as Britain and France constructed their drug control regimes in the early twentieth century. Anxieties about the practical effects that an unregulated domestic opiate market might have on the economic health of the nation prompted tighter narcotics controls in Britain: it was when the open availability of the drugs began threatening the nation's economic and commercial well-being that the state stepped in to regulate them more stringently. In France, however, the state took action when fears that opiates could affect social and political solidarity were realized with the spread of drug use in the military. Each nation's rationale for controlling opiates continued to drive policy into the 1920s, when the question of how the state should treat its addicted citizens came to the fore. In Britain, authorities realized that opiate addiction often was not inimical to economic productivity or self-sufficiency, and they decided to allow some addicts to receive maintenance doses so that they could continue leading productive lives even while regularly injecting drugs. In France, on the other hand, narcotics still posed a threat that was as much ideological as it was practical, and the authorities did not devise a system for maintenance treatment or provide resources to cure people of their addictions. Throughout the period, thinking about what constituted the national community—and what sort of threat opiates posed to it—framed and drove the development of drug policies in Britain and France.

Beyond differences between British and French anti-narcotic nationalisms, the other major difference between the drug problems faced by the two nations was demographic. In Britain, nonmedical opiate use seemed to be on the decline throughout the late nineteenth and early twentieth centuries, and many British addicts were able to continue functioning as normal, contributing members of society in spite of their drug habits. Given that addiction was rare and seemed to have few adverse social consequences, the spread of opiate habits, though still worrisome, was not imminently threatening for the British. In France, however, addicts were believed to be both more numerous and more socioeconomically diverse than their British counterparts, and the consequences of their drug habits captured the public eye. France had its share of addicts whose drug habits did not cause much social harm, yet opiate use seemed to be on the rise in the late nineteenth century, with commentators decrying its spread across the socioeconomic spectrum, from industrial workers to artists, socialites, and soldiers. Thus, whereas the British addiction problem was seemingly contained within the private sphere and not causing much disturbance, French drug habits seemed to be growing and to have the potential to cause problems throughout society, from the factory to the high-society salon and the military barracks.

In the late nineteenth and early twentieth centuries, then, the British and French faced not one unified drug *problem* but rather two distinct drug *problems* that seemed both qualitatively and quantitatively different. In Britain, opiate habits were seen as threats to the self-sufficient and capitalist ethos of the nation, but the dangers seemed to be waning rather than growing; in France, opiates represented a menace that was not only practical but also ideological, and the problem was apparently on the rise. Thus the threats in the two countries were of a different character and on a different scale, leading the authorities to tackle their drug problems in dramatically different ways.

From Drug Discourses to Drug Policies

This book covers the development of drug discourses and policies in Britain and France from 1821 through 1926. The start date, 1821, is the year that Thomas De Quincey first published his "Confessions of an English Opium Eater," a work that established how the opiate experience would be understood for much of the next century. The end date, 1926, is when the differences between British and French approaches to the drug problem became clearest, with publication of the British Ministry of Health's report on opiate addiction, which established that maintenance treatment was an acceptable way to care for addicts. The initial steps to control opiates in Britain and France were originally taken with the passage of pharmaceutical regulations in the mid nineteenth century, but I address these developments only in passing because they are already discussed at length elsewhere. My policy focus here is on the early-twentieth-century regulations that centered on narcotic drugs in particular, since these were, as historian Virginia Berridge says, "the first legislative measures to establish narcotics as a matter of social policy."[18]

I begin, in chapter 1, by discussing two characteristics of opiates that made them distinctive in the nineteenth-century mind: their internalizing qualities and their addictive nature. Citing philosophical works, medical research, and popular representations of opiate use, I outline how the experience of using opiates came to be seen as fundamentally different from the nineteenth century's most common form of recreational intoxication—alcoholic inebriation. Alcohol had long been considered a social substance in the European imagination, whereas in the nineteenth century, thanks in large part to the writings of De Quincey, opium developed a reputation as an elitist and antisocial intoxicant. Opium was seen as a drug that represented a voluntary and powerful retreat away from the social and toward the interior, thus making it the chemical embodiment of social disengagement. Beyond opiates' facilitating an experience that was subjectively different from other forms of intoxication, nineteenth-century researchers also recognized that the drugs were addictive. Often, they found that opiate addiction was not just a disease or disorder but an indicator of broader character faults. In some cases, opiate habits were understood as symptoms of excessive individualism, while in others, they were just seen as instigators of selfish and haughty patterns of thought. Consequently, opiate addiction was at times understood as a physiological metaphor for its victims' personalities—a medical manifestation of self-exploration and disengagement run amok.

In chapter 2, I move beyond general understandings of what the opiate experience meant to a discussion of fears about its potential social and political consequences. I begin with an overview of the scope and scale of opiate habits in the two nations in the late nineteenth century, showing that while opiate use seemed to be on the decline in Britain, it was spreading—and apparently starting to cause more harm—in France. Beyond these quantitative differences, there was also a qualitative divergence in the sort of threats opiate habits represented to the national communities of Britain and France. Each country's brand of anti-narcotic nationalism was colored by a broader sense of national identity and citizenship, and the anxieties articulated in writings about opiates expressed concern about what the drugs could do to each nation's residents and their ability to act as good citizens. In Britain, fears centered on worries that opiates could transform upstanding Britons into indolent, lazy individuals tainted by their association with an "Oriental" drug. In France, concerns about the effects of opiates on the spiritual and ideological health of the community were expressed through stories describing the link between opiate use and treason. Around the turn of the century, two well-publicized and emblematic examinations of opium as it related to these questions came to different conclusions. The findings of an 1895 Royal Commission helped defuse British worries about the effects of opiate use on self-sufficiency and the fulfillment of social roles, thus somewhat mitigating the threat that opium represented to the British ideal of citizenship. In France, however, the 1907 treason trial of Charles Benjamin Ullmo reinforced, rather than relaxed, the connections between opiate use and poor citizenship.[19] By 1910, therefore, concerns about both the significance and the prevalence of opiate habits were becoming less pressing in Britain but more so in France.

In chapters 3 and 4, I analyze the development of drug policies in Britain and France in the first three decades of the twentieth century. Chapter 3 focuses on the beginnings of narcotics control, telling the story of how and why opiates were eventually governed under a separate control regime and were no longer treated as pharmaceuticals. In Britain, despite the concerns of a few vocal medical men, reformers, and local authorities that domestic opiate use was problematic, officials in Whitehall took little notice of the issue until the summer of 1916. Only when the open availability of the drugs began to threaten what Britons held most dear—commerce—did the government take action. The issues that lax controls over opiates were causing had little to do with domestic consumption by addicts, as the policies were more directly tied to the problems that England-based smuggling operations were causing with Britain's trading partners. To cut down on smuggling, the government decided (at the behest of a shipping company) to cut off, at the source, access to narcotics that were being smuggled overseas and to make it more difficult for international smugglers to get provisions on the British mainland. Thus it was concern about the nation's financial health, not the physical health of individual drug users, that pushed the state to action. British anti-narcotic discourse in the nineteenth century and narcotics control policy in the twentieth century were thus intimately related, both driven by fears of what the unrestricted flow of opiates could do to the country's economic well-being.

The link between nineteenth-century anti-narcotic nationalism and twentieth-century narcotics control policy was also manifest in France. Military authorities began working to restrict access to opiates to keep the drugs out of soldiers' hands, particularly since opium use among servicemen who traveled to Indochina was a constant problem. In October 1908, the government took action, placing limits on the freedom to buy, sell, and use opiates, not just for soldiers, but for the entire population. With the outbreak of World War I, the opiate question once again emerged as a military concern. By 1915, the French authorities started expelling individuals involved with drugs from combat zones and interning them in prison camps that were otherwise reserved for Germans, Austrians, antimilitarists, and other perceived threats to the war effort. The isolation of drug users from the national community was no mere contingency of wartime, as subsequent regulations allowed the authorities to seize the property of drug law offenders, strip them of their civil liberties for up to five years, and expel them from their home cities for up to ten years. Thus the Republic that proclaimed itself the home of "liberty, equality, and fraternity" sanctioned the incarceration, impoverishment, and isolation of individuals who violated drug control regulations. In short, if a Frenchman was caught up in narcotics, he could be stripped of the rights and privileges of citizenship in the Republic. Given the prevailing understanding of the opiate experience as one that, by its very nature, was antithetical to the French conception of good citizenship, such harsh responses to those who indulged in drugs were hardly surprising.

Finally, in chapter 4, I analyze the British and French approaches to the treatment of opiate addiction. By the 1920s, addicted patients and the doctors who treated them recognized the dilemmas posed by control, and some advocated for changes in policy to allow addicts to continue leading relatively normal, productive, and socially integrated lives while still using narcotics. Confronted with similar problems, authorities in London and Paris took drastically different paths to finding a solution. In Britain, the Home Office solicited the Ministry of Health for advice, and in 1926, the Departmental Committee on Morphine and Heroin Addiction issued a report advising that in intractable cases of addiction, physicians could stabilize their addicted patients on low doses of opiates if they needed the drugs to function.[20] Addiction and good citizenship, as understood in Britain, were not irreconcilable, and the state made allowances for the maintenance treatment of patients if their doctors believed they could lead better and more productive lives with narcotics than without them. No such concessions were made in France at this time, and the response of the state was to handle addiction not by providing better treatment options but by undertaking a more enthusiastic prophylactic strategy to limit the availability of narcotics. Changes in the British and French drug-using populations also contributed to the differences in the two nations' attitudes. By the 1920s, the drug-using population in Britain seemed to contemporaries to be both small and benign, whereas in France it was apparently becoming larger, more varied, and more socially menacing in its behavior. Discursive and demographic factors merged in the 1920s, leading to a divergence of British and French policies on addiction.

The size and nature of each country's drug-using population played a role in shaping approaches to drug control and addiction, but conceptions of citizenship also influenced the trajectory of British and French anti-narcotic nationalisms and policies. Opiate control and addiction became one of many issues in which abstract ideas about society merged with the perceived realities of a public problem to steer the directions of both discourse and state action.[21] That the nation with the more virulent anti-narcotic nationalism (France) also seemed to have a more serious drug problem helped broaden, rather than narrow, the gap that was emerging between British and French attitudes to drug use and addiction. Thus, with their divergent understandings of both the sociopolitical significance of opiate use and its spread across their societies, by the 1920s, Britain and France had developed two parallel approaches to the drug problem. Just as Boekhout van Solinge found in his analysis of European drug policies since the 1970s, both the social characteristics of drug users and the relationship between states and their citizens influenced British and French approaches to handling addiction when drug control first became a matter of public policy, at the beginning of the twentieth century.[22] Thus British and French drug discourses and policies were not just responses to a common set of health dilemmas caused by a psychoactive substance. Rather, they were a social and political set of solutions to the challenges posed by drugs that were considered not only physiologically and psychologically harmful but also potentially socially and politically poisonous.

CHAPTER ONE

Imagining the Meditative Nation

Constructing the Opium Experience

Man, in all ages, has shown a frantic taste for all substances, benign or
dangerous, which exalt his personality and testify to his grandeur.
—*Charles Baudelaire*

In the nineteenth century, Britons and Frenchmen became increasingly aware
of the powers, both enjoyable and excruciating, of psychoactive substances
and began to explore them in greater detail. While more focused on alcohol,
the substance most prominent in their societies, poets, medical researchers,
and commentators also began to discuss the nature of opium, creating a new
way of thinking about its physical and psychological effects. Opiate intoxication and addiction, though understood to have a good deal in common with
drunkenness and alcoholism, also came to be seen as qualitatively different
from them. Writers in Britain and France saw the experience of being under opium's sway as more than just an alternative form of drunkenness; to
take opium was to join a "meditative nation" separate from the lives of one's
friends, neighbors, and fellow citizens. Opiate consumption, even among
those who began using the drugs with no intention of experiencing their
psychoactive effects, came to be seen as a potentially antisocial practice with
implications that were not just personal and medical but also political.

The Experience of Intoxication

During the eighteenth and nineteenth centuries, Europeans became increasingly curious about psychoactive drugs and intoxication. In part, this

was due to changes in drug-using practices, as heavy alcohol drinking became more prominent among the working classes and excessive opium use
was more widely publicized. Cultural and intellectual currents of the era also
played a role in making Europeans more interested in understanding the nature of psychoactive substances and the possible consequences, both good
and bad, of their use. Intellectuals were intrigued by the mind-expanding
potential of drugs, and moralists and psychiatrists found alcohol and opiates
to be intriguing objects of study because of their effects on both physical and
mental functioning. Together, by the end of the nineteenth century, these
developments merged to reframe how Europeans viewed drinking and drug
habits.

Alcohol and the People

If every civilization has its poison, Europe's is alcohol. Despite moral
denunciations of drunkenness throughout the centuries, alcohol staked its
claim as the principal psychoactive substance of the Judeo-Christian world.
In the Middle Ages and the Renaissance, alcohol was ubiquitous in daily life,
as beer and wine provided readily available alternatives to disease-carrying
water. The nutritional value of beer made it a staple of the European diet, and
wine was a key ingredient in medicines and folk remedies, since it was believed to ward off illness. The drinking of fermented beverages found its way
into a variety of carnivals, festivals, and rituals, linking alcohol consumption
not only with individual intoxication but also with populist merrymaking
and sociability.[1]

In time, the links between alcohol consumption and camaraderie became
embedded in the cultural world of the masses in western Europe. The most
prominent of the social rites surrounding alcohol was the toast, which dated
back to the eleventh century. Another tradition that emerged was buying
rounds of drinks for friends, which transformed alcoholic libations into gifts.
Sharing a pitcher or bottle became a well-understood gesture of goodwill, an
excuse for strangers to strike up a conversation, or a way for old acquaintances
to rekindle the bonds of friendship. Even heavy drinking was often seen as a
quaint and charming excess, as most popular representations of intoxication
painted the drunkard as a character who was affable, goofy, and loveable,
more than anything else. In the eighteenth and nineteenth centuries, the connections between alcohol and sociability were cemented with the evolution
of public drinking spaces, which became prime centers for people to both

drink and socialize. Watering holes evolved into places where friends and fellow workers could gather and strangers could meet and engage in casual discussion, political debate, or even courtly pursuits. With the introduction of the modern bar in the 1820s, drinking establishments became welcoming alternatives to the restrictive private spaces of genteel society. Thus alcoholic beverages provided the social lubricant, and drinking spaces became the loci of social intercourse, places where friendship, camaraderie, and political awareness could germinate and flourish.[2]

As both facilitators of and excuses for sociability, alcoholic beverages came to be associated not just with the local neighborhood but with the larger community of the nation. Social drinking unleashed what Charles Baudelaire termed "the music of patriotic passion." For the British, the drink of choice was beer, which along with Protestantism and constitutionalism completed a trinity of Britishness, in contrast to the wine, Catholicism, and autocracy of the French. John Bull traditionally carried an ale pot in his hand, and as Daniel Defoe wrote of the "true born Englishman," the "mob are statesmen, and their statesmen sots." For the French, wine was the drink of the Republic, a beverage that, as a popular saying went, "wore no breeches." As one nineteenth-century physician argued, wine contributed to the "cordiality, frankness, and gaiety" that differentiated the French from their beer-guzzling neighbors in Britain and Germany. Thus, just as it united the popular classes with bonds of sociability at the local level, alcohol also connected citizens within the larger collectivity of the nation. Drink and solidarity went hand in hand, and to consume alcoholic beverages was not just to be a good friend and neighbor; it was also to be a good citizen. As anthropologist Marianna Adler writes, "*not* to drink was tantamount to a complete withdrawal from socially meaningful existence as it was then defined." On both an individual and national plane, alcohol truly was the poison of the people.[3]

Despite the place of alcohol in the daily lives of the many, there were a vocal few who argued that drunkenness had a dark side, as alcoholic inebriation could easily slip from an enjoyable and loose state to a mad and dangerous one. As long as drinking had been a custom in Europe, there were commentators who objected to it, equating it with blasphemy, waste, moral degradation, disruptive behavior, and revolutionary political ideas. A primary example of this can be found in Thomas Nashe's *Pierce Penilesse*, published in 1592. According to Nashe, there were several different "species" of drunk, most of which corresponded to an animal: the "ape drunk" would leap and sing to his heart's content; the "lion drunk" was a violent beast; the "swine drunk" had an insatiable appetite; the "sheep drunk" sat in a sullen stupor; the "goat drunk" was a lecherous predator; and the "fox drunk" was a mischievous and crafty creature. Tracts decrying the evils of alcohol some two centuries later continued to depict the drunkard as an animal, warning readers that strong drink could "render people more like beasts than men." Drunkenness, warned writers concerned with the consequences of excessive drinking, was apt to bring out humans' less civilized side, unleashing a tide of imbecilic behavior, poor impulse control, and violence.[4]

By the late nineteenth century, researchers had developed more precise ideas concerning the nature of alcoholic drunkenness, postulating that the physiological effects of alcohol led drinkers into states of excitement, stimulation, and unruly behavior. Some doctors held that alcohol's effects stemmed from its ability to coagulate fluids and cause reactions in the blood that led to increased nervous activity. Others maintained that it acted as a stimulant, altering the hemoglobin in the blood or somehow acting directly on nerves. Given that alcohol seemed to stimulate drinkers at first, only to sedate them later, theorists understood drunkenness as more of a multistage process than one unified state. First, researchers believed, drinkers would become excited, happy, and garrulous, before giving way to decreased inhibitions, extreme emotions, poor muscle coordination, and a weakening of the intellectual and moral faculties. Finally, in the highest state of drunkenness, they would succumb to what British addiction specialist Norman Kerr termed "paralysis," entering a virtually comatose state in which they were completely unresponsive to outside stimuli.[5]

Consequently, though alcohol was in many respects the social substance of nineteenth-century Europe, some recognized that alcoholic inebriation could have decidedly antisocial effects, making drinkers more isolated from and hostile to others. Commentators began to recognize this dark side of alcohol, particularly after the emergence of concerns about drink leading to public disorder, poor work habits, and crime among the industrial working classes. Furthermore, with the spread of new, more potent forms of alcohol—such as gin in Britain and alcohols distilled from beets and grains in France—some began to fear that excessive indulgence in alcohol could cause irreparable damage, both physical and moral.[6]

Outside the temperance fringes that emerged in Victorian Britain and fin-de-siècle France, however, most commentators maintained that if the right types of alcohol were consumed in the right amounts, drinking could remain an acceptable social and recreational activity. As Kerr explained, indulgence in alcohol (and the resulting drunkenness) could have either a social or an antisocial character, depending on who was drinking and for what reasons. The "social" drinker, he elaborated, was a man who drank "openly and without disguise, and rarely, except in congenial companionship with other drinkers." For this type of drinker, alcohol was truly a social substance: it was "to his fondness for company that he is indebted for his introduction to drinking." But, Kerr warned, there was a potential for drinking to become a "solitary" practice, one in which the drinker "shuns the company of his fellows, shuts himself up with the poison . . . and chews the cud of his morose imaginings or indulges in vain dreams." For these individuals, alcohol could transmute from a social substance into an isolator, one that separated them from friends and neighbors instead of facilitating interaction with them.[7]

Thus, although alcohol was generally seen as a social substance, commentators in the mid to late nineteenth century began to recognize its potential to instill in the heart and mind of the drinker what medical writer A. T. Shearman termed "a state of distrust and dislike for his fellow creatures" and an "anti-social feeling." At certain stages of drunkenness and for certain drinkers, alcoholic beverages could become the opposite of social tonics, leading the drunkard to become a misanthrope, someone who failed to fulfill his responsibilities to those around him, or both. Yet alcohol's antisocial potential was generally considered an unwelcome side effect of drunkenness only for the minority, not a constitutive part of drinking for the majority.[8] Compared with alcohol, other substances would develop a much stronger reputation for being inimical to social interaction and belonging, not just when used to excess, but whenever their psychoactive effects took hold.

A New Pastime on the Horizon: Opium

Along with alcohol, another substance figured prominently in European medicine by the late eighteenth century: opium. Extracted from the poppy plant *Papaver somniferum*, opium was drawn out by slitting the heads of the poppies and then underwent a process of boiling and drying before being smoked or "eaten" in the form of pills or liquid preparations. The substance had a long-established place in Europe, going back to the Greek physician Galen, who found its analgesic powers to be useful in the treatment of problems ranging from deafness, epilepsy, and kidney stones to minor ills such as headaches, coughs, and menstrual pains. During the Crusades, Christian knights learned more about opium, observing how Arab peoples used it to treat dysentery, diarrhea, and eye disease. In the Middle Ages it became one of the commodities exchanged in the Venetian trade, and in spite of attempts to grow poppies domestically, Europeans found that foreign opium was more powerful and cost-effective. The trade of opium grown in India, Persia, Egypt, and the Levant up through Italy, France, Germany, and England became significant over the medieval and early-modern periods.[9]

As it became more readily available, opium evolved into a mainstay of the European pharmacopoeia. Demand for the substance increased throughout the early-modern period, particularly after Paracelsus concocted his laudanum tonic: a mixture of opium with henbane, pearls, coral, cow intestines, amber, musk, and crushed seashells—an elixir that he claimed was superior to the heroic remedies of the time. Indeed, compared with many medical procedures of the day such as bloodletting, leeches, blistering, and large doses of laxatives and mercurials, opiate tonics were effective and pleasant medicines that lacked the complications associated with traditional treatments. In the seventeenth century, doctors developed more precise laudanum recipes, the most popular being those perfected by Englishman Thomas Sydenham, who concocted a mixture of opium, saffron, cinnamon, cloves, and wine, and by Rousseau, a French monk, who combined opium with apple juice, nutmeg, saffron, and yeast. With laudanum, opium came to be seen as one of the most powerful treatments available in Western medicine, and its use was nearly ubiquitous by the late eighteenth century. Beyond laudanum, opium also began appearing in various teas, lozenges, and capsules prepared by pharmacists, grocers, salesmen, and quacks. In the age before aspirin, opium became the panacea of choice, providing relief for everything from hangovers to body aches, asthma, and cancer. As one medical dictionary advised in 1801, "there is scarcely any disorder in which, under circumstances, its use is not found proper."[10]

Though in agreement that opium was highly effective, medical men could not reach a consensus on why it worked so well. According to Galen, opium was a "cold" substance, one that gained its power from its ability to induce sleep. Others considered it a "hot" substance that operated by distending and rarifying tissue as it warmed the body. Later theories became more mechanical in their explanations, as researchers postulated that opium acted as either a stimulant or a sedative. On this question, however, there was contradictory evidence: observations showed that when first ingested, the drug increased the pulse, only to suddenly shift course and make the entire body operate more slowly and laboriously than before. As researchers were unable to resolve these contradictions, more elaborate explanations emerged. Some argued that opium worked by altering the density of the blood and nervous fluids, instead of simply slowing their circulation; others postulated that opium particles were shaped like hooks or arrows, which stuck to nervous membranes and obstructed the flow of fluid from nerves to the brain. Despite the emergence of a voluminous literature on the subject, opium remained a mystery to the medical practitioners who put it to use. Of all medical substances, British physician Samuel Crumpe wrote in 1793, "there is none whose nature has been less understood, or with respect to which a greater diversity of sentiment is entertained."[11]

Even while marveling at its great healing powers, doctors recognized that if used injudiciously, opium could have unpleasant effects. British physician George Young, for example, took a large dose of laudanum in 1753 for a severe cough. In short time, Young reported, he began experiencing a loud ringing in his ears, confusion, and a loss of feeling in his legs; frightened by these symptoms, he forced himself to vomit and bled himself to purge the drug from his system. Yet, as Europeans had seen in their trips abroad, the psychoactive effects of opium were not unpleasant for everybody. In the Middle East, where opium eating was a popular pastime, and in China, where the substance was smoked, visitors observed the drug's effects with a combination of horror and fascination. Travelers generally assumed that foreign peoples used opium in the same way that wine and beer were used recreationally at home—to "procure gayness and pleasant inebriety." Just as they knew alcohol could lower inhibitions, increase sociability, and sometimes lead to boorish behavior, Europeans believed that opium could do the same for the peoples of Asia and the Middle East. Generally, however, observers tended to focus on the more spectacular effects of opium intoxication, figuring that since non-Europeans had "smaller brains" and more delicate nervous systems, they were less likely to enjoy loose conversation and convivial merrymaking when intoxicated. Reports from abroad, such as Frenchman Baron de Tott's 1785 travel memoir from the Middle East, focused on how Turks would gesticulate and jabber nonsense when they took opium. Captain Cook's description of Malays' opium use asserted that they became prone to wanton violence and murdering sprees when under its influence.[12]

The travel and medical literature of the eighteenth century defined recreational opium use as an Oriental practice, but the psychotropic effects of the substance were certainly experienced by Europeans as well. Chaucer and Shakespeare made brief references to opium's ability to induce drowsiness

and sleep, and Thomas Wharton Jr., writing in 1747, spoke of the drug's capacity for creating "dreams of wanton folly." Eighteenth-century writings
about opium pointed out how some Europeans overdosed on the substance
"by design" with the hope of procuring its exhilarating effects, and researchers who experimented with the drug noted that it was able to produce a "gayness" akin to alcoholic drunkenness. By the nineteenth century, opium taking had evolved into a pastime for some British and French workers, and it
was particularly prevalent among common folk in the English Fens. In most
cases, the practice seems to have been regarded as an adjunct or alternative
to recreational drinking.[13] In Romantic circles, however, the psychoactive effects of opium would soon take on a new dimension that went beyond the
alcoholic experience.

The Pope of the Poppies

Views of opium intoxication changed in large part because of cultural
currents of the late eighteenth and early nineteenth centuries. Romantic
writers of the era craved a means for finding "transcendental subjectivity"
and accessing previously unexplored realms of the imagination. In particular, many sought what literary historian Marcus Boon terms the "flash of
gnosis," a transmission from the fallen and corrupt material universe into an
"inner world" where the truth of the human spirit resided. By overcoming
boundaries imposed by earthly limits of time, space, and the body, gnosis
was a means for achieving a greater understanding of what lay beyond the
senses and concealed in the hidden realms of the mind. One way of procuring
this flash of knowledge, according to Novalis and other leading Romantics,
was through mind-altering substances that could give the mind free rein by
eliminating awareness of the body and the environment from consciousness.
Some used hashish, ether, or alcohol to these ends, but literary experiments
using opium were particularly promising; soon Romantics came to believe
that the drug was a key to the liberation of the mind from physical limits. As
Samuel Taylor Coleridge described in his letters to Thomas Wedgwood at the
end of the eighteenth century, opium had the potential to unlock the world
of the sublime for the philosophically inclined: a "new secret brotherhood,"
as Coleridge put it, could develop around opium, one in which philosophers
would voluntarily poison themselves in order to procure gnosis through the
substance's effects.[14]

Coleridge also hinted at his belief in the special power of opium in the preface to *Kubla Khan* in 1816. Coleridge wrote that the poem came to him while he was under the influence of opium and fell into a deep sleep; when he awoke, he found several hundred lines of poetry composed on the paper before him. While he was unconscious, opium took possession of him and, seemingly, wrote the poem, rich in Oriental imagery and fantasy. Opium, Coleridge claimed, authored words and images that would not have flowed from his pen if he was in a state of normal consciousness. Linking him to foreign lands and extraordinary powers of expression that came out in the poem, opium seemed to provide a transcendental passageway for Coleridge, one that allowed him to overcome normal limits of distance and vocabulary. Coleridge offered no explanation of why or how opium acted on his mind and poems as it did; such an explanation would come a few years later with the work of one of his disciples, Thomas De Quincey.[15]

De Quincey undertook a detailed study of the effects of opium, not merely mentioning the substance in passing, but devoting a major body of philosophical work to elucidating what opium intoxication was like and how it functioned. He began his explorations in 1821 with his "Confessions of an English Opium Eater," which was published in parts in *London Magazine* before being issued as an independent volume the following year. His study continued in the 1840s, with his "Suspiria de Profundis" (1845) and "The English Mail Coach" (1849). De Quincey's opium explorations were written in the first person, based, he claimed, on his own experiences with opium from the first time he consumed it, in 1804, and through the rest of his life. Like most people of his time, De Quincey did not originally come to opium for recreational purposes. He accidentally discovered its intoxicating potential when a friend suggested he take laudanum for a rheumatic toothache. Once he began consuming it regularly, however, it became a pastime for him, and its pleasures differed drastically from the experience of consuming opium simply to alleviate physical pain.[16]

According to De Quincey, opium did not gain its powers by *creating* dreams, visions, or transcendence; instead, it served as a key that unlocked the hidden recesses of the brain and freed buried sentiments and visions in the mind. The human brain, De Quincey believed, was like a palimpsest, a vellum scroll that was capable of being erased and rewritten upon, but always maintained faint traces of all the words ever inscribed upon it. Just as, with the application of proper chemicals, past inscriptions on a palimpsest could become clear, De Quincey held that unknown but important impressions on the mind could be brought into focus with opium. Thus the drug was more than a simple intoxicant; it was a means to access the interior, a substance that had the ability to uncover parts of the imagination that were normally hidden from consciousness due to distance or time.[17]

Thanks to its power to help transcend temporal boundaries, De Quincey often saw long-forgotten visions from his childhood when he took opium. Beyond experiences from his own life, opium also ushered De Quincey into a deeper and more profound past, transporting him to scenes of ancient and "Oriental" fantasy, inundating him with a hodgepodge of bizarre animals, ancient temples, and foreign gods, spanning from ancient Egypt to China. Though exotic, these Oriental visions were not just forms of escape but also a type of internal exploration. As literary historian Barry Milligan points out, in De Quincey's era the Orient represented both the foreign and an "all encompassing origin" that contained the roots of humankind buried in the distant past. De Quincey shared this perception of the Orient, seeing it as a place where "antediluvian man renewed" still existed. His visions of a stereotypical "East," therefore, were not simply a dreamlike trance that brought him to foreign places; they were a result of opium working its natural processes of intellectual excavation. Just as opium helped the individual remember forgotten scenes from his past, it also revived the deepest roots of human consciousness, which, in De Quincey's thinking, lay in the mystical East. Thus the Orient, as Milligan explains, was "at once both other and origin" for De Quincey, and his opiate-fueled voyages to the deepest recesses of the human spirit revealed the most ancient scripts on the palimpsest of the human brain—those of its Oriental roots. Consequently, by transporting the opium user to faraway times and places, the drug simultaneously facilitated a voyage inward, as the deepest levels of human consciousness and awareness ultimately lay in the "ancient" and the Eastern. To take opium was, for De Quincey, to take a journey outward from contemporary Western society, but to do so by way of the interior, ultimately giving the opium user access to an Orient that was, paradoxically, both exotic and primordial.[18]

According to De Quincey, what gave opium its power to help transcend the limits of time and space was its ability to open up awareness of sights, sounds, and sensations that were not limited by the eye's ability to see, the ear's ability to hear, or the skin's ability to feel. The drug, De Quincey wrote, helped one reconnect with "long-buried beauties" of the past and build "cities and temples" out of the "fantastic imagery of the brain," while bringing oblivion to whatever earthly problems preoccupied the user. The outside world, therefore, was ultimately of little import for the opium user, as the drug's unique powers lay in its ability to privilege internal visions, dreams, and memories over the here and now. Thus opium was able to help man transcend what De Quincey termed the normal "limits of any human experience" and achieve gnosis, with his mind running free and unfettered by the limits on the senses and consciousness that were normally imposed by the body.[19]

In this respect, opium was particularly valuable for De Quincey: it held the promise to serve as a chemical corrective for what he saw as shortcomings in the social, cultural, and intellectual life of his time. Modern existence, he believed, had dulled the imaginative faculty in many, as men had become too concerned with the physical and the social to explore the inner strata of their minds. For De Quincey, developments such as steam power and the daguerreotype were not modern marvels that improved humanity's lot but rather "yokes" that made "daily experience incompatible with much elevation of thought." Objects were not the only culprits in the decay of the meditative faculty, however; excessive engagement with others was also guilty of fostering intellectual stagnation. "Living too constantly in varied company," De Quincey wrote, left thought "dissipated and squandered." If one led "too intense a life of the *social* instincts, none suffers more than the power of dreaming." Thus to lead a truly philosophical and intellectual life, the scholar needed to free himself from the physical and social distractions that tied him to the world outside his mind. Opium proved a perfect means for rebelling against the tyranny of the tangible and the bonds of excessive sociability, as it possessed what De Quincey termed a "*specific* power" to enhance thought and dreams while drowning out distractions from the outside.[20]

Though opium could stimulate the mind, De Quincey maintained that the drug was not for everyone. The positive effects of opium, he believed, could only be as powerful as the intellectual capacity of the individual who consumed it. When taken by an ordinary man of average imagination or intelligence, he explained, opium would have rather unspectacular effects. "If a man whose talk is of oxen should become an Opium Eater, the probability is, that (if he is not too dull to dream at all) he will dream about oxen." Similarly, the positive effects of opium were probably lost on the "barbarians" who lived outside Europe, since they were not "capable of any pleasures approaching to the intellectual ones of an Englishman." What enabled De Quincey to fully appreciate opium's effects, he wrote, was that he was intellectually superior to most others. Throughout his work, De Quincey emphasized his contemplative and philosophical nature, and how he was above the petty distractions of physical comfort and sociability that dominated the lives of the masses. Instead of simply relying on his senses to color existence, De Quincey considered himself a man in "possession of a superb intellect in its *analytic* functions" and superior "*moral* faculties" that gave him a unique "inner eye and power of intuition of the vision and mysteries of our human nature." Endowed with such a unique mind and temperament, De Quincey considered himself among the few with the capacity for truly appreciating, understanding, and elucidating what opium could do. Thus he considered himself the "alpha and the omega" of the "true church on the subject of opium," a "Pope . . . and self-appointed *legate a latere* to all degrees of latitude and longitude." As an intellectual being who considered himself above and beyond the norms of sociability and a sensual life, De Quincey believed that he, and a select meditative few, were the only ones equipped to appreciate and utilize the full potential of opium.[21]

To emphasize how opium—and opium intoxication—was not for everyone, De Quincey took pains to describe how the drug's effects were vastly different from the intoxication known to most people of his time: that procured by alcohol. "Opium," he explained, "is incapable of producing any state of body at all resembling that which is produced in alcohol; and not in *degree* . . . but even in *kind*." Opium differed from alcohol not simply in the "quantity of its effects merely, but in the quality." Alcohol, De Quincey believed, gave drinkers a fleeting pleasure that privileged the "merely human, too often the brutal" part of man. Drink transformed a man into a simple and "sensual creature" who would rely on the physical senses to experience the world and was more likely to socialize than to contemplate. Consequently, alcohol would "volatize and . . . disperse the intellectual energies," leading the drinker to the "brink of absurdity and extravagance." Opium, by contrast, gave a "steady and equable glow" to the mind, one that helped man explore the "diviner part of his nature" by bringing a "great light" to his "majestic intellect." Thus opium had the power to "concentrate what had been distracted" by the social and modern. Whereas alcohol privileged the senses and society over the self, opium brought the mind into focus, giving it freedom to roam and exercise its powers without limitations or distractions that could disturb intellectual exploration and meditation.[22]

Thus alcohol and opium were, for De Quincey, diametric opposites. Alcohol brought the drinker out into the world of physical stimulation and sociability, and opium brought the user deeper into his own mind, on an adventure of self-excavation that would be compromised by interacting with the outside world. To drink alcohol was to experience crass stimulation, whereas

to eat opium was to undergo a deeper intellectual experience that strengthened connections not with others but with the self. Thus, while the drinker
was a good friend and comrade, the opium eater was in a class apart.

Providing one of the first detailed examinations of recreational opium use
from a European perspective, De Quincey set the trajectory for one of the discourses on the drug that would develop over the next century. As he claimed
he would, De Quincey became one of opium's main spokesmen. Within two
years of publication of his "Confessions," fifteen reviews came out, expos-
ing the work to a wide audience. In time, De Quincey would also become
a popular icon of recreational opium use in France, gaining notoriety after
Baudelaire's summary of his work in *Les paradis artificiels* in 1860. A hundred
years after De Quincey's works first appeared, medical writings about opiate
addiction would still use the "Confessions" as a case study, and as late as the
1920s, some addicts would name De Quincey as the man who inspired them
to try drugs. Of course, De Quincey probably inspired only a small percentage of those who began using opiates, as the vast majority of the opiate-using
population first started using drugs for therapeutic, not philosophical, reasons. Nonetheless, the durability of De Quincey's account provided one of
the foundations of how Europeans would regard recreational opium use and
helped define its significance within the popular imagination.[23]

De Quincey's Legacy: Opium, the Interior, the Egoist, and the Orient

Three strains of De Quincey's thought on the psychoactive effects of
opium remained prominent through the beginning of the twentieth century.
First, his emphasis on how opium thrust users toward the interior, making
them more concerned with their own minds than with the outside world
or with others, shaped the way that writers—both medical and literary—
envisioned the opium experience as distinct from alcoholic drunkenness.[24]
Second, De Quincey established a link between opium use and elitism with
his assertion that it took an intellectual predisposition to enjoy the drug's
pleasures "properly." Third, with his emphasis on the Oriental visions and
fantasies that opium could procure, De Quincey strengthened the preexisting
link between recreational opium use and the foreign.

In the generations after De Quincey, theorists remained unclear on the exact mechanisms that made opium work, though there was a paradigm shift in
understandings of how the drug operated. In the eighteenth century, researchers considered opium, like alcohol, to be a substance that acted on the lower

regions of the body by altering the nervous and circulatory systems, but by the nineteenth century experts shifted the focus of their opium studies from below the neck to the area above it. Writers came to see opium as a substance that acted not as an anesthetic or a relaxant for nerves and muscles but as a drug that gained its powers over the senses by acting on the brain itself. Above all, opium was now seen as a more "cerebral poison" than alcohol, one that affected sensory perception by altering how outside stimuli were interpreted. According to British researcher Mordecai C. Cooke, this effect was due not to a distortion of the senses themselves but rather to a hyperactivity of the mind when under opium's sway. "The senses convey no false impression to the brain," he explained, noting that the nervous system seemed unaffected by the drug. "All that is seen, heard, or felt, is faithfully delineated, but the imagination clothes each object in its own fanciful garb." Though the opium user could see a normal object, and its image would be transferred from the eye to the brain without distortion, the mind under opium's influence was prone to "exaggerate . . . multiply" and "color" it beyond recognition. It was the mind's interpretation of outside stimuli, not their reception or transmission to the brain, that was altered by opium.[25]

With the interpretation of sense perceptions left to the whims of the imagination when an individual came under the influence of opium, the mundane would become fantastic. The brain would misinterpret signals from the eyes, ears, skin, and tongue and would experience sights, sounds, touches, and tastes that might or might not be related to any actual stimulus. But soon after the poison took hold, researchers believed, the brain would become hypersensitive, and all sensations would become extremely intense and uncomfortable. Even the slightest stimulation—a light, a sound, or a touch—at this point could cause unpleasant sensations for the opium user. Consequently, at this stage of intoxication, the person would seek isolation and shelter from the senses. Hypersensitivity would then fade about an hour later, leading the brain's perception of outside stimuli to become "attenuated" and "blunted." Physical sensations would soon be reduced to a minimum in the user's brain, as senses became "progressively numb" and gradually displaced by visions, sounds, and sensations coming from the hyperactive mind. Thus the sensory windows connecting individuals with the external world would gradually shut over the course of an opium episode. At first, stimuli would seem distorted due to the drug's effects on the brain, only to become muted and eventually leave the user almost completely disconnected from the goings-on around him.[26]

Researchers believed that the isolation of the opium user from the outside world reached its pinnacle when he entered the stage of "somnolence," in which all signals to the brain would "completely disappear" and a "sensorial veil" would separate him from his surroundings. As British physician Robert Little wrote of the opium user in this state, "objects could hit the eye, but they are not seen; sound may fall upon the ear, but none are heard," and soon the user would be engulfed in a state described by contemporaries as one of "comatose insensibility." Describing this experience based on his own experiments with the drug in 1877, French author Charles Richet explained that at this stage of opium intoxication, he felt that "the exterior world" would simply "disappear," making it seem that "there [was] no longer anything but an interior world" centered in his mind. Even the opium user's body language indicated that he was shutting out the external in a rush toward the internal core of the mind. As French researcher Bénédict Morel observed, the body would seemingly "collapse in on itself," mirroring the opium-induced mental and spiritual retreat to the interior.[27]

As opium freed the mind from outside distractions, researchers believed, it had the power to "liberate intelligence from its normal chains." Sensual isolation would be complemented by mental hyperactivity, an excited imagination, a sharpened memory, and an ability to think with "singular lucidity." "The intellectual excitation provoked by opium," French psychiatrist Roger Dupouy wrote, "precipitated the procession of thought, multiplied the association of ideas . . . put mental representations in greater relief," and "amplified . . . syllogistic faculties," producing an intellectual experience far more profound than an ordinary hallucination.[28]

Many writers maintained that the cerebral stimulation facilitated by opium was directly related to the drug's ability to limit, or eliminate, the influence of physical and social distractions on the thought process. "Liberating intelligence" from the usual state of "corporeal servitude" that subsumed the brain's activity to what the physical allowed, opium seemingly set thought free from the bounds of the body. Many commentators went as far as to describe opium-induced reveries as more of an out-of-body experience than a mere intoxication. Researchers who took opium experimentally noted that when under its influence, they felt a sense of isolation from the physical and a sense that "one could no longer feel their body, and becomes all thought," with an intellect that seemed "removed from the head." To be under the influence of opium was to become "an intellect, a *pure* intellect," one capable of experiencing joys much more profound than any "felicity of flesh" or anything in the tangible world. To take opium, according to French author Claude Far-rère, was to enter a new realm, becoming "less carnal, less human," and "less earthly." Set free of all physical limits, the opium user would fancy himself, as Farrère described, "superhuman." In this respect, opium seemed to provide the gnostic flash that Coleridge and the Romantics sought, freeing the intel-lect and the imagination by unbridling them from their restricting physical bonds.[29]

Concomitant with the apparent "disembodiment" of the intellect was the loss of physical restraints of time and space, just as De Quincey suggested. Poets and researchers noted how opium had the power to seemingly subvert the normal givens of time, producing dreams in which individuals traveled millions of years in a single vision, and witnessed events from the ancient past and the distant future simultaneously. Opium would make space seem infi-nite, "without bottom or limits," enabling users to "fly far above the earth and far above the theater of human strife." Granting access to the infinite, opium seemingly procured an experience that, for British physician George G. Sigmond, was "indescribable" and "almost supernatural." Thus to engage in opium use was not only to escape the restrictions imposed by the physical and social bodies but to truly transcend and soar above and beyond them on a higher plane of existence.[30]

Consequently, opium intoxication came to be seen not simply as more in-tense than alcoholic drunkenness but as qualitatively different. Theorists held that while alcohol's psychoactive effects arose from its physiological action, with opium, mental disruptions were not a mere side effect of hyperactivity or physical damage to organs or nerves. Opium's action was predominantly intellectual, whereas alcohol's was fundamentally "mechanical," since it af-fected the body more than the mind. "Intoxication with opium is nothing like intoxication with alcohol," French author Albert de Pouvourville noted, "no more than an intellectual is like an animal, since the former satisfies curiosities of the spirit, while the other gratifies the excessive appetites of the brute." An author writing in the *Lancet* came to similar conclusions, stating that whereas "wine seem[s] to invigorate the animal frame, opium [invigo-rates] the intellectual powers." When an individual fell under the influence of opium, explained a doctor at a meeting of the Westminster Medical Society in 1839, he would sit as "still as a marble ... apparently in the deepest reverie," in contrast to the "quarrelsome, bustling, and noisy" drinker. Thus opium was not a substance that led to outbursts or distortions because of hyperstimula-tion, as alcohol was believed to do, but one that completely altered how the human organism operated—cutting off connections with the outside world and leaving the user alone with the psychotropic caprices of his own mind. The drug did not simply blur how the world was experienced by users, it fundamentally changed it.[31]

Researchers believed that since opium launched the user into a completely cerebral sphere, its effects would vary, depending on the individual's personal and intellectual makeup. Like De Quincey, subsequent writers emphasized that the drug did not *create* dreams, visions, or insights, but rather brought them out from the hidden recesses of the mind. Since opium reveries would reflect the thoughts, and often the desires, of the user, they were often painted with brushes of self-aggrandizement, facilitating visions of fulfilling personal hopes and fantasies. According to researchers, opium dreams were generally "megalomaniacal," especially since the drug made users believe they had gained heretofore undiscovered insights. Feelings of superiority, self-satisfaction, pride, self-love, and a sense of being divine were commonly reported psychoactive effects of opium. As Farrère's narrator confessed in *Black Opium*, when under the drug's sway, the "genius" brought about by opium created a sense of being "the best . . . among men." "Surely the Olympus of the Greeks and the Paradise of the Christians have no blessedness in store for their elect. And yet, such are my blessings," he crooned. "From the first puffs . . . I become obviously the superior of men."[32]

As late as the 1920s, such understandings of opiates' psychoactive effects on the ego remained prominent in both Britain and France. Aleister Crowley wrote in 1922 that opiates' ability to transform a human into a "god descending upon earth" helped him transcend the simple ideas that governed ordinary men. Even for those who suffered the tortures of addiction, such as French writer Jean Cocteau, opium's ability to transform a person into a "masterpiece" made it well worth taking. To tell an opium user that he was "degrading" himself with the drug, he quipped, was akin to "telling a piece of marble that it has been deteriorated through Michael Angelo, or a piece of canvas that it has been stained by Raphael, a piece of paper that it has been soiled by Shakespeare, or silence that it has been broken by Bach."[33]

Alongside the egotism engendered by opium, elitism developed around the question of who was capable of utilizing the drug's psychoactive properties "properly" and who was not. Echoing De Quincey's admonition that opium could not create visions but only bring out what was hidden in the mind, nineteenth-century authors argued that only a select few could utilize the drug to truly gratifying ends. If people took opium with the hope of achieving "sensations, instead of ideas," they warned, they would be disappointed. The drug was best used by "a few intelligent and conscientious spirits," since, as Dupouy explained, "reverie with opium is an intellectual act that not everyone knows how to do or is capable of achieving." The philosophical use of opium was an "art" to be practiced only by those with minds and temperaments capable of engaging it with the right mindset.[34]

Since indulgence in opium for its psychoactive properties was understood as a means for isolating users from their physical and social surroundings, the practice came to be seen as a mark of individualism and snobbery. Both in its effects and in how it was to be used, the drug tended to encourage condescension, reverie, and isolation rather than fraternity, camaraderie, and interaction. Researchers who compared opium and alcohol use noted this difference, highlighting that whereas the majority of drinking was done with friends or family, opium taking was "rarely a social act," and the opium aficionado preferred to take the drug "in private" whenever possible. As historian Terry M. Parssinen correctly points out, "unlike the noisy, convivial British pub, the opium den was a place where each smoker was alone with his pipe and his dream . . . when one went to a den, one did not escape *to* society . . . but *from* society." In France, Farrère's writings about his opium experience also gave credence to the idea that the opium user sought isolation from his social world. When taking opium "I no longer worry about anything," he wrote. "I no longer have any trade and friends . . . opium . . . plunges me deeper within myself. And I find it sufficiently interesting to enable me to forget all that is outside." Entranced by the enchantments afforded by their own minds, recreational users seemed to consider themselves, as literary historian Alethea Hayter concludes, "isolated, both elect and outcast," superior to and apart from their peers. And as historian Virginia Berridge has shown, opium's ability to provide a chemical shortcut to "denying society" was recognized in late-nineteenth-century Britain, leading some who were seeking to "retreat into the individual" and escape the "vulgar materialism of the external world" to turn to the drug.[35]

Thus to indulge in opium was to enter, as Baudelaire called it, "an appalling marriage of man to himself" that made users too enamored with the psychedelic to engage in the social. Overly focused on the happenings in his opiate-affected brain, the user could become subject to what Laurent Tailhade termed a self-imposed "excommunication from . . . the society of man." Though opium was still mostly being used as a medicine and a palliative agent in the nineteenth century, it also came to be seen as "the ultimate flower

of voluntary and lucid alienation" for those who used it with the express purpose of enjoying its psychoactive effects. As such, opium, when recreationally
used, was not simply an intoxicant that, like alcohol, could ease nerves at
social gatherings; it was an isolator and an insulator that built walls between
individuals rather than promoting sociability.[36]

Opium's reputation as an antisocial intoxicant was amplified by its association with the foreign. Whereas beer, wine, and spirits were well entrenched
in national traditions in Britain and France, recreational opium use was traditionally seen as an Oriental indulgence. In part, this was because the effects
of opium seemed to have affinities with the "Oriental character." Because it
engendered placidity and a dreamlike state, many writers believed that opium
matched the "spirit of life in the Far East," which according to the prejudices
of the time was marked by a "supreme joy of non-activity" and contemplation; as writers on both sides of the English Channel maintained, "the ideal
of Whites" was "activity," whereas the "ideal of the Oriental" was "inertia . . .
passivity" and "beatific oblivion." Consequently, opium, with its tendency to
produce torpor and nonphysical stimulation, seemed to suit the "Oriental"
character quite well, while representing the antithesis of the "White" ideal.[37]

Historical circumstances gave European writers fodder for the argument
that recreational opium use was an Oriental—and more specifically, Chinese—practice. In large part, this was due to the actions of Europeans themselves. In the late eighteenth century, the British began an illicit opium trade
with China, shipping opium grown in India to the Middle Kingdom through
a system of smuggling and bribery. In a short time, the clandestine trade
became a lifeblood of the British East India Company, as the sale of Indian
opium in China compensated for commercial deficits elsewhere and kept
British opium manufacture profitable. Despite the Chinese government's efforts to stem the opium trade, it became more firmly institutionalized as a
result of the two "Opium Wars" (1839–1842 and 1856–1860), in which the British compelled the Chinese to open up to foreign commerce and accept the
opium trade on their soil. The subsequent growth in the opium trade saw a
seventyfold increase in Chinese opium consumption from the 1760s through
the 1850s. The French also played a part in spreading the habit in Asia, as they
established an opium monopoly in Indochina and made a handsome profit
selling the drug to their colonial subjects, particularly those who had emi-
grated from China. Due to the size of the opium business in Asia, by the end
of the nineteenth century, commentators in both Britain and France came to

consider opium the "national poison of China" and the "ethnic poison of the yellows."[38]

Oblivious to the geopolitical and economic dynamics that fortified the opium habit in East Asia, many Europeans thought that the prevalence of habitual opium use in China stemmed from faults in Chinese society and culture. Stereotypical writings generally assumed that since life in East Asia was boring and Asians were intellectually inferior, Asian peoples used opium to procure some slight pleasures in an otherwise depressing existence. The drug seemingly had the ability to sooth "John Chinaman into a temporary forgetfulness of his troubles," appeasing him with visions of pleasure. More often than not, speculations of what these reveries were like were less than flattering and sometimes shockingly racist, even for the times. In the opium dreams of the Chinese, speculated one British commentator in 1863, lay "a Paradise where the puppy-dogs and rats run about ready-roasted . . . where the men all have short names and the women short feet, where everybody . . . has a right to order everybody else three hundred strokes of the bamboo." Just as opium pleased the philosophical European with thoughts of transcendence, it would tantalize the Oriental with images that reflected the allegedly "backward" spirit of the East.[39]

The confluence of imperialist facts and stereotypical fictions helped color the image of opium to the point that the drug became the very essence of the Orient in the European imagination. According to some writers, opium had the power to act as a "magician" that could usher European users into states of Asian consciousness. When opium was consumed, they wrote, it had the power to substitute Chinese blood for hardy and "thick" European blood, providing a means to "penetrate" the Asian mindset. In some accounts, long-term use could even transform a European into an Asian. French writer Daniel Borys, for example, described meeting an opium user in Marseille who was no longer recognizably French: he had developed a "bilious complexion, circumflexed eyelids, curbed eyes, and a hanging mustache" that made him look more Chinese than European. British authors also considered opium a substance that could transform ethnicity, writing that the drug was prone to turn the "whites" of one's eyes into "yellows." Thus to indulge in opium recreationally was, as literary historian Susan Zieger concludes, to engage in a "particularly corporeal and performative mode of Orientalism," an act of simultaneously consuming and embodying the exotic.[40]

Opium, therefore, was not only a drug that could isolate users by cutting off their consciousness of the outside world and their desire to interact with it, but a substance that also could segregate Britons and Frenchmen from their compatriots in a more obvious and tangible way—by altering their identity. In the nineteenth century, recreational opium use was marked as foreign and exotic, making the practice an alien and unnatural one in the Western imagination. When combined with the antisocial and isolating character that, according to contemporary writers, the drug could foster, recreational opium use evolved into the chemical embodiment of snobbery, antisocial tendencies, and alienation all rolled into one. Compared with the association of moderate alcohol use and occasional drunkenness with fraternity and national identity, opium's face was one of egotism and foreignness; whereas alcohol was a poison of the people on both the personal and collective levels, opium drew distinct lines between users and those around them. Thus in the discourses that emerged around opium in the nineteenth century, opium users—or at least those who indulged in the drug to enjoy its psychoactive effects—seemed self-absorbed outsiders rather than outgoing members of their local and national communities.

The Dark Side of Psychoactive Substances: Addiction

Despite its palliative, recreational, and philosophical uses, there was more to opium than just escape. While having the power to liberate the individual from physical pain and sensorial boundaries, opiates also could enslave him in the body, creating bonds of dependence that, by the late nineteenth century, would be recognized as a physical and behavioral disorder.

Addiction, broadly understood as dependence on the use of psychoactive substances, took many forms, as medical researchers in the second half of the nineteenth century gradually recognized that alcohol, opiates, and other drugs all had the potential to become habit-forming. By the end of the century, many within the medical community believed that addiction, to substances ranging from caffeinated beverages to tobacco, alcohol, and opiates, was part of one overarching disorder. In Britain and the United States, leading physicians who specialized in the treatment of alcoholics and drug addicts termed this disorder *inebriety*, which they defined as a "diseased condition, the characteristic symptom of which is an overpowering impulse to indulge in intoxication at all risks." In France, similar understandings of the disorder they termed *toxicomanie* emerged in the late nineteenth century, though some considered alcoholism to be distinct and separate from problems caused by excessive recourse to other psychoactive substances. Individuals with nervous disorders, a constitutional predisposition to disease, and tainted heredity, the proponents of inebriety theory argued, were particularly susceptible to the charms of psychoactive substances and were prone to turn to them to relieve physical or emotional stress. After becoming initiated to their drug of choice, medical theorists argued, inebriates would do permanent pathological damage to their nerves, necessitating repeated and ever-increasing doses to achieve the desired psychoactive effects.[41]

Yet, even as researchers recognized that addiction had many common mechanisms and causes, they also noted that certain kinds of people were more likely to abuse certain kinds of substances, and they classified different substances and addictions accordingly.

Alcoholism

Europeans became most familiar with the potential problems caused by excessive indulgence in psychoactive substances by studying the one they knew best: alcohol. A voluminous medical literature on alcohol-related problems emerged in the late eighteenth and early nineteenth centuries that focused more on the consequences of prolonged alcohol abuse than on the habit-forming nature of the substance, addressing correlations between heavy alcohol use and liver abnormalities, delirium tremens, and rotten teeth. The term *alcoholism* first emerged in 1849 with the publication of Swedish researcher Magnus Huss's *Alcoolismus Chronicus*, which was quickly summarized in medical journals across Europe and by the early 1850s had set the standard for research on alcohol and its effects. Subsequent European studies of alcohol elaborated and expanded on the many complications enumerated by Huss: physical symptoms including loss of appetite, trembling hands, hallucinations, trembling tongue, loss of motor function, stomach pain, cirrhosis, lesions on the brain and internal organs, fatty degenerations of cells, and pathological changes to blood vessels; and psychological complications ranging from jealousy and hypochondria to hallucinations, criminal impulses, insanity, and suicidality. Many of these damages to the mind and body were considered permanent, and according to some writers, prolonged abuse of alcohol could not only cause complications but kick-start the degeneration—or "reverse evolution"—of drinkers and their offspring.[42]

Beyond examining the problems caused by prolonged and heavy alcohol use, European researchers in the nineteenth century also began to recognize the addictive nature of the substance. Until this time, compulsive recourse to drink was not considered a symptom of a disease or disorder but was seen as a form of gluttony—an inability to restrain oneself from indulging in a sensual pleasure. Gradually, however, some observers began to note that heavy drinkers were enslaved by drink rather than simply indulgent in it. In 1784, American physician Benjamin Rush was among the first researchers to draw a direct correlation between drinking and dependence on alcohol, observing that while drinking was originally a choice, it could quickly devolve into a habit and then a necessity. Gradually, a degree of medical consensus emerged that the morbid desire for drink could be caused by excessive alcohol consumption. "Alcohol . . . produces by habitual use a craving for itself," British psychiatrist Frederic C. Coley explained in 1904. As "continued use induces resistance," the drinker required "ever-increasing" amounts of alcohol to achieve the intoxication he sought. According to the editors of the *Lancet,* writing in 1887, through an unknown "pathological process, alcohol began inebriety." This craving was thus recognized as not so much an indulgence as an "urgent compulsion" to drink. Consequently, alcoholism was not a "selfish fault" but a disorder, something that the drinker was powerless to stop.[43]

The question of what made one person an alcoholic and another a moderate drinker, however, remained a mystery to most doctors. The breadth of alcohol-induced problems observed by researchers made it hard to understand, or even define, alcoholism. In part, this was because the drinkers that doctors treated did not have a common, identifiable set of symptoms; some were violent, others passive, others confused and disoriented. Unable to completely explain alcoholism, many doctors began to eschew purely clinical explanations for chronic drinking problems and insert other, more character-oriented factors into the equation. Though they acknowledged that the drinker had a physical need for alcohol, many experts on alcoholism pointed an accusing finger at the drinker's will, arguing that it was a desire for pleasurable intoxication driven by a "weak character" or poor willpower—not just an involuntary impulse—that led individuals to drink themselves sick. Some, like British physician Patrick Hehir, unabashedly drew a correlation between the "moral considerations" and the "psychical problems" that the drinking habit brought to the fore. "It is not merely . . . physical necessity" that led the drinker to overdo the amount of alcohol consumed, he argued, but a "debased moral nature" that thirsted for the psychoactive effects of drink. Alcohol, Hehir concluded, was a serum that revealed the "original basis of the character" rather than itself facilitating excess. The correlation between moral failing and physical degradation was thus appropriate. "Habitual drunkenness is not a disease, it is a vice," Hehir charged, comparing the drinker to the "corpulent glutton" who brought ill health on himself with his uncontrolled appetite. Alcoholism was not the culprit in the drinker's decline; rather, it was a fitting end for the drinker's lack of restraint. Other researchers, while not going as far as Hehir, echoed his sentiments, warning that though excessive alcohol use was "sometimes a disease," practitioners should not fall into the blunder of "supposing that it is always so."[44]

Researchers conceded that a moral tinge stained the scientific study of alcoholism, noting that many theories of chronic drinking were "based on the theory that the disease is only sin." While recognized as a disease, alcoholism also became a nosological metaphor for what was seen as a lack in character: an inability to refrain from excess in the pleasurable experience of getting drunk. Alcohol not only caused problems for the drinker but also highlighted deficiencies within him. Thus, the argument went, alcoholism was often a natural and just fate for its victims.[45]

The Opium Habit

Europeans also recognized that opium could be habit-forming, diagnosing uncontrolled use first among foreign peoples, then among regular users closer to home. The habit-forming nature of opium had been recognized as far back as Roman times, and Westerners witnessed the ravages of addiction first-hand in their travels to the Middle East. In 1582, for example, a Spanish physician traveling in the Ottoman Empire noted that, if not careful, opium eaters could kill themselves with the drug, and that prisoners already accustomed to using opium could perish if not given doses while incarcerated. Travelers in Persia observed similar phenomena, even going as far as terming the opium habit a "disease." As opium became more popular as a medicine in Europe, even its most ardent proponents also understood its dangers. Sydenham recommended that his laudanum tonic not be used too regularly, and others observed that individuals who used the drug for a prolonged period were prone to start taking it in "uncommon large quantities." Nonetheless, most medical writers before the nineteenth century agreed that opium's analgesic benefits far outweighed its potential drawbacks, and habitual recourse to opium was generally seen as a normal indulgence and elicited little comment.[46]

As poets and artists began using opium toward "philosophical" ends and documenting their experiences, the dangers of opium addiction became more evident. Coleridge and George Crabbe fell victim to the opium habit around the end of the eighteenth century, making the potential problems of opium ever clearer to those who also extolled its philosophical and spiritual virtues. Among the most insistent on the drug's potential pitfalls was De Quincey himself. He devoted an entire section of his "Confessions" to what he termed the "pains of opium," telling of his perpetually increasing doses of laudanum and the problems they caused him. In the throes of dependence, De Quincey found, opium tended to have "palsying effects" on his intellect rather than stimulating it. "Opium had long ceased to found its empire on spells of plea-sure," he admitted, bitterly commenting that "it was solely by the tortures connected with the attempt to abjure it that it kept its hold." De Quincey advised his readers to "fear and tremble" at the altar of opium, and though he still recommended it, he warned that a strong will and self-discipline were needed to avoid the pitfalls of dependence. In the decades after the "Confes-sions," medical researchers in both Britain and France concurred, warning that opium could be "imperiously addictive" and that, once habituated, users were prone to consume massive amounts of the drug.[47]

Even more striking than the testimony of De Quincey and some domestic commentators were the reports coming back to Europe from the Far East, where, by the mid nineteenth century, medical men had begun sounding alarm over what they observed among opium-consuming populations. In particular, an 1850 article by Little caught the attention of the medical com-munities in Britain and France and became one of the seminal texts in the ar-gument against opium use for the rest of the century. Little studied the effects of opium by examining the owners of opium shops in Singapore, the smokers who frequented the establishments, prisoners who used the drug regularly, and paupers supported by local poor houses. Among the main consequences of opium use, Little found, were physical problems that ravaged the body from head to toe: insomnia, headaches, difficulty breathing, excessive secre-tions of mucus from the nose and eyes, constipation, discharges from the genitals, body aches, and a loss of muscle tone. As the habit progressed, users could not eat or drink unless they were under the influence of the drug, and their bodies would reach a degraded state that left them unable to resist dis-ease or infection.[48]

Even worse than the effects of opium, however, were the consequences of abandoning it once habituated. Little described despondency, exhaustion, diarrhea, and vomiting that would accompany withdrawal but could be relieved with a fresh dose of opium. Given the drug's ability to soothe the pains of withdrawal, it was highly habit-forming: "No one who has once fairly given himself up to this unhappy vice will surrender it voluntarily." Little found that opium users would spend all their money on the drug, neglecting their wives and children and resorting to crime to feed their morbid appetites. Once an individual picked up the opium habit, therefore, he became ensnared in a trap: either succumb to the physical effects of excessive consumption or face the torturous and potentially lethal consequences of going without it. Whichever he chose, the ultimate result would be the same: physical decay followed by a rapid descent into social disrepute and an early death. As Morel grimly noted in his summary of Little's research, "if you know when an individual started smoking opium, it is easy to predict when he will die. His days are numbered."[49]

In the fifty years after Little's work, travelers and medical experts on both sides of the English Channel became increasingly cognizant of the physical dangers of opium for habitual users. A body of literature emerged describing how opium would gradually intoxicate the body, inflicting physical (insomnia, muscle pains, convulsions, dry skin, vomiting, heart palpitations, sterility), mental (nightmares, diminished intelligence), and moral (apathy, moroseness, laziness) troubles as users increased their dosage to procure the drug's euphoric effects. Though at first people might use it casually, medical writers warned, the drug could quickly become "an imperious need." Researchers maintained that if taken to excess, opium could eventually kill the user by asphyxiation, by coagulating the blood to the point that it could no longer circulate properly, or by flattening the ridges on the brain so that it no longer functioned properly. "The devil could not have invented a more pernicious vice for the destruction of soul and body than this of opium," a British observer warned in an 1894 testimony before the British Royal Commission on Opium, asserting that the drug had the power to transform the habitual user "into the most forlorn creature that treads the Earth."[50]

Commentators maintained that as the body accustomed itself to the poison, the opium user would require ever-increasing doses, to the point of consuming enormous quantities. Even more frightening was that excessive opium use seemed inevitable for a large number of those who dabbled in the drug. It took months of heavy drinking to become an alcoholic, but researchers believed that the opium habit could be formed in a matter of days. "It is even more true of opium than of alcohol," the *Lancet* editors warned in 1887, "that no middle course is possible between abstinence and infatuation."

Thus, though it was possible to be a moderate drinker, it was difficult to avoid addiction with opium. Addiction to opium was also exceedingly difficult to cure. As British doctor I. Pidduck lamented, "it is no more possible to reform a person who has been long addicted to the practice of taking opium, than it is to reform a patient who is paralyzed."[51]

Opium's Heir: Morphine

Opium clearly posed a threat to some users, but the proliferation of morphine, an opium alkaloid, created a new and much more prevalent addictive menace in Europe. Morphine was introduced into European medicine in the early nineteenth century. In 1825, German pharmacist Heinrich Emanuel Merck began wholesale production of the drug, and by the 1830s it was common enough to earn a place in most European pharmacopoeias. Doctors found the new drug helpful in the treatment of asthma, bronchitis, diarrhea, dysentery, insomnia, malaria, arthritis, rheumatism, sciatica, herpes, toothaches, migraines, nervous angina, and secondary syphilis, and a magnificent calmant. The main complication associated with early morphine treatments, which were usually taken orally, was that they induced vomiting. However, medical providers found that injecting morphine helped avoid emetic complications, and with the popularization of the hypodermic syringe in the 1850s, injection became the preferred route for administering the drug. Soon, morphine became a new wonder drug in European medicine, part of what German psychiatrist Edward Levinstein termed the "modern method" for relieving pain. By the 1870s, it had proven so effective that doctors began giving it in almost all situations where patients were suffering, even for mild discomfort or anxiety. As British practitioner Clifford Allbutt remarked in 1870, morphine had become the hallmark of a new generation of medicine, making "the syringe and the phial . . . as constant companions" to contemporary doctors "as was the lancet to their fathers."[52]

In dispensing morphine so liberally, doctors also vulgarized knowledge of the drug and its powers, and some patients began experimenting with it on their own to cope with everyday problems. By the 1870s, psychiatrists claimed that not only the sickly and the incapacitated were using the drug but so were statesmen, artists, and doctors. Contemporaries began observing that morphine was even becoming a "fashion" in certain circles, as prostitutes, intellectuals, and members of the upper classes used it to procure "new thrills" that traditional recreational substances such as alcohol could not provide.

This was particularly the case in France. In 1881, *Le Temps* took it for granted that morphine was becoming a "well-known" fashion in European nightlife, and commentators would soon speak of morphine as a growing "epidemic" among social elites. Journalists observed the evolution of Parisian "morphine institutes," upper-class gatherings that were a combination of traditional salons and modern-day "shooting galleries." Recourse to morphine, according to the editors of the *Lancet*, was becoming an "idler's pastime," and like the drive to use opium recreationally, it seemed to result from a desire to experience pleasant dreams and visions that could not be experienced without the drug.[53]

As the use of morphine, as both medicine and recreational pastime, proliferated in the 1870s and 1880s, many researchers began to understand the drug's psychoactive effects as being similar to those of opium. They believed that, once injected, morphine gave an initial sense of cerebral overstimulation, marked by vertigo, insomnia, distortion of tactile senses, abnormal physical reflexes, hallucinations, and excitability. These sensations then gave way to an attenuation of the senses, followed by feelings of physical pleasure. Doctors maintained that these pleasant experiences did not actually come from the body but were ruses of the mind, which shaped and distorted the signals it received from the body. At the height of intoxication, morphine, like opium, seemingly enveloped the consciousness, isolating the user both physically and emotionally from those around him and dissipating connections with the outside world. Adrift in what French psychiatrist Claudius Gaudry termed an "ocean of egoism," the morphine user seemingly became a "misanthrope" as he fell further into the habit, an individual who was "fiercely apathetic" to friends and family and was concerned with "nothing but himself." Prone to a form of what French physician Daniel Jouet called "intellectual decadence," individuals who sought out morphine for its psychoactive effects were believed to be subject to a "moral downfall" akin to that believed to accompany recreational opium use.[54]

Increasing familiarity with morphine also gave doctors an opportunity to study its potential drawbacks more closely, and it did not take them long to learn that even more than opium, morphine could be extremely addicting. In December 1870, Allbutt was one of the first in Europe to publish an article on morphine's potential to become habit-forming. While acknowledging that morphine injections were of great value in the practice of medicine, Allbutt noticed that patients who grew accustomed to morphine began experiencing discomfort when they went without it. Thus, he warned, the drug had the potential to "induce in those who use it constantly, an artificial state which makes its further use a necessity." And because fresh doses of morphine seemed to alleviate the discomforts of withdrawal, the drug had the potential to quickly become an "artificial want" akin to that kindled by the opium habit.[55]

By the middle of the 1870s, researchers in continental Europe began to confirm Allbutt's suspicions. The first comprehensive work on the morphine habit came from Levinstein, superintendent of the Schoenberg-Berlin Asylum, with a presentation in Berlin in 1875 and publication of a monograph, *Morphiumsucht*, in 1877. Levinstein defined the "morbid craving for morphia" as "the uncontrollable desire of a person to use morphia as a stimulant and tonic, and the diseased state of the system caused by the injudicious use of said remedy." Morphine, Levinstein warned, had the potential to induce physiological and psychological changes "in the entire system" when ingested regularly, transforming habitual users into patients for whom the norms of medical and psychiatric practice would not always apply. In short time, *Morphiumsucht* was summarized, translated into English and French, and tweaked by medical minds throughout Europe, helping make the study of morphine addiction its own subdiscipline within psychiatry by the early 1880s.[56] As the drug became popular in Britain, and even more so in France, doctors, psychiatrists, and sociologists worked throughout the last two decades of the nineteenth century to refine the framework laid out by Levinstein into a more cohesive and thorough theory of morphine addiction, its mechanisms, and its etiology.

In late-nineteenth- and early-twentieth-century studies, doctors held that one thing that distinguished morphine addiction from the opium habit was how quickly it ensnared its victims. Once morphine users were psychologically implanted with a desire to regain the psychotropic paradise they had experienced with their initial injection, researchers believed, they could be drawn back to the drug after just one taste of its cerebral stimulation. The window of opportunity to escape the allures of morphine was brief, and after a few doses, many users would succumb to temptation and begin taking the drug regularly. Soon thereafter, tolerance would set in, making the injection of morphine a "bizarre, strange" and "tyrannical" habit that users turned to when faced with the slightest physical pain or emotional struggle. When the transitory effects of the drug passed, bodily and mental depression would return even stronger than before, tempting the morphine user to inject once more. Soon, doctors predicted, the user would be so accustomed to having the drug in his body that he could not function without it; in its absence, he would become somber and morose, but a fresh injection would transform him into a light-hearted and affable creature. After a sufficient number of injections, researchers warned, morphine would become essential at nearly all times, rapidly leading users toward constant intoxication and extraordinarily high doses. Consequently, morphine was extremely habit-forming. According to many medical experts, it was, unlike alcohol, capable of catching all who tried it in the snare of dependence. Although alcohol could be enjoyed in moderation, "everyone . . . who takes morphine regularly," warned Gaudry, "becomes a morphine addict."[57]

In the schema laid out by late-nineteenth-century researchers, morphine habits inevitably led addicts down a road to physical and mental demise. Hoping to procure the hallucinatory effects that small doses once gave them, users would inject themselves with massive amounts of the drug that could kill them through cardiac arrest, gastric problems, circulatory disorders, or suffocation from respiratory complications. Even if they did not fatally poison themselves, doctors warned, habitual users doomed themselves to a miserable existence as the morphine habit would wear down their physical health. Addicts would age prematurely, with hollowing eyes, yellow or ash-colored skin, shivers, nausea, constipation, impotence, and wrinkling of the skin that transformed them into living cadavers. In such a state, doctors warned, addicts would waste away, with their systems becoming so weak that a minor complication or disease that a healthy individual could survive might prove fatal. Morphine would also exhaust the brain, as intellectual torpor would inevitably replace the resplendent visions once produced by the drug, and psychological disturbances ranging from hallucinations to suicidal thoughts would haunt the morphine taker.[58]

Beyond all this, the drug was also believed to transform habitual users into monsters who were proud, rather than ashamed, of their addictions. Researchers believed that unlike compulsive drinkers, who were often embarrassed by their drunken behavior and would bitterly regret their binges, morphine users reveled in their excesses. Doctors maintained that addicts developed a fetishistic love of morphine, becoming obsessed not only with the drug itself but also with the instruments associated with injection. As they grew accustomed to the habit, addicts were alleged to become morbidly fascinated by the shine of their needles, and a regularly mentioned symptom of morphine addiction was users' tendency to compulsively clean and organize their hypodermic collections. In France, wealthy addicts were rumored to order syringes made of gold and silver that were enameled, covered with

emblems, engravings, monograms, and jewels, both personalizing and sanc-
tifying the tools they used to inject. Many were believed to even derive a perverse joy from injecting themselves in body parts that, in the demure words
of French psychiatrist Benjamin Ball, "would make us cringe with disgust."
Thus, unlike drunkards, morphine addicts drew what doctors deemed a perverse sense of pride from their habits. Consequently, researchers maintained,
most morphine addicts did not want to be cured; the drug, it seemed, con-
quered not only the bodies of the addicted but also their souls.[59]

According to practitioners who specialized in the study of the morphine
habit, the perverse pride derived by addicts was not necessarily a result of the
intellectual complications that drug use engendered. Even though doctors
recognized that morphine could cause insanity, many did not believe that
the drug debilitated the brain. Like opium, morphine was believed to stimulate the mind, not dull it. Some suggested that the drug actually increased
intelligence and made users think more lucidly. "There is no mental disease
showing itself during the poisonous action," nor "the least detriment to . . .
capacities" among regular morphine users, Levinstein explained. A "rapid
change of temper," rather than "mental alienation," was morphine's main
psychological effect. Levinstein maintained that morphine addicts were both
in "full possession of their intellectual faculties" and capable of rising to the
foremost ranks of science, art, the military, and medicine, even as they took
the drug habitually. Morphine did not incapacitate the brain but seemingly
made it work with a fiendish alacrity, not altering its ability to function but
shifting the motives that drove thoughts and actions. The drug's main effect
was to "make the moral axis deviate," with the will and ethical standards,
but not the capacity to reason, being thrown into disarray by prolonged use.
By specifically targeting and disabling the brain's moral capacities, medical
researchers believed, morphine left the addict morally bankrupt yet still capable and, consequently, potentially dangerous as well.[60]

Opiate Addiction: Disease or Vice?

Given that opiates had both medical and recreational uses, it was often
difficult for late-nineteenth-century practitioners to ascertain why and how
their patients with substance abuse problems had become addicted. If, for
example, a morphine addict came to a doctor for help after starting to take
the drug for intellectual stimulation and pleasurable sensations, would it be
logical to treat the disorder in the same way as that of an addict who started

injecting morphine to assuage the physical pains caused by a terminal illness? Could the drug habits of those addicts who sought psychoactive pleasure and those who had turned to opiates for pain relief be considered one and the same?

A very small minority of opiate addicts who sought treatment in the late nineteenth century, and thus wound up serving as case studies in the medical literature, first took opiates for their pleasurable or philosophical psychoactive effects. The vast majority were iatrogenic addicts, individuals who started using opiates as part of treatment for a chronic disorder or for pain relief, then were unable to stop when their organic ills had dissipated. In the 1870s, most medical practitioners agreed that it was irresponsible prescribing practices, along with naivety about opiates' habit-forming potential, that allowed addiction to develop in these instances. In cases of iatrogenic addiction, the doctor was to blame for the onset of opiate addiction, since, as Levinstein explained, the prescribing physician was the "originator and propagator of the disease." Pharmacists, the argument went, were also guilty, as they failed to recognize customers' addictive behaviors and did not refuse to fill prescriptions for large amounts of opiates over an extended period.[61]

Yet by the 1880s, as physicians became more cognizant of the dangers inherent in prescribing opiates, the onus of addiction shifted decidedly to the addict himself. Medical writers maintained that as doctors became aware of the dangers of addiction, they began to dispense opiates more carefully, and for a briefer duration, to prevent the development of drug habits. But for those who became addicted, even a brief therapeutic dose or regimen was enough to give them a taste for opiates' psychotropic pleasures, and they began using trickery and deception to procure the drugs even when they no longer needed them. Whether by malingering, making up new pains, or claiming they became sick when they started to taper their doses, researchers believed, addicts would con their doctors into giving them prescriptions that were medically unnecessary, thus further feeding and reinforcing the habit. By the time doctors recognized that these patients had become full-blown addicts, it could be too late. Medical researchers wrote that once addicted, patients would begin to lie, forge prescriptions, or steal to continue procuring supplies of drugs. Thus, even though many cases of addiction may have had their roots in the doctor's office, it was the patient who kept pushing for more drugs, not the doctor who compassionately prescribed them, who was to blame for the development of substance use disorders. "Once exclusively coming from therapeutic origins, morphine addicts . . . can no longer use the old excuse of medical influence," French practitioner Georges Pichon charged. The editors of the *Lancet* agreed, casting blame for addiction on the addicted themselves, since based on common knowledge and warnings from doctors, "they know the danger which lurks in their newly found pleasure." By the 1890s, according to some writers, this made the excuse of iatrogenic addiction untenable. "Today there are no longer any who are poisoned," French doctor Henri Guimbail wrote in 1891, "only those who become intoxicated with morphine." The addict, therefore, was "hardly deserving of pity," since he had only his own appetites, his lack of self-control, and the series of lies he told to continue procuring drugs to blame for his addiction.[62]

According to many addiction theorists, addicts' character shortcomings were not necessarily the result of intellectual complications engendered by opium or morphine use. Rather, the drugs served as what Guimbail called a "provocative agent," one that may have helped highlight but did not necessarily create users' antisocial or immoral streaks. Opiates, he explained, "only develop latent tendencies," bringing out "hidden instincts and secret aptitudes" rather than creating them.[63]

Many in the medical community echoed these sentiments, claiming that there were certain types of individuals with certain "latent tendencies" who were drawn to opiates' psychoactive charms. The first type were victims of heredity, "psycho-chemically deficient" or "unbalanced" in their nervous systems and unable to produce the cerebral stimulation that created healthy thoughts and senses of pleasure. To compensate, these individuals were naturally attracted to opiates, which could provide a shortcut to stimulation by simulating normal neural activity. The second type of addiction-prone individuals were those with poor self-control, or a "disease of the will," as medical theorists termed it. After initially experiencing opiates' charms, such individuals lacked the discipline necessary to resist them. The third type were the intellectually and philosophically inclined who suffered from boredom and turned to opiates for amusement, to break the monotony of the everyday. An overly "cerebral life," a "poetical temperament," or a "bizarre" personality created a particular receptivity to the intellectual stimulation and isolation that opiates provided. The intellectual strata of society were believed to be naturally drawn to the drugs specifically because of their psychotropic qualities. Just as "a butterfly will flap his wings near the light of a lamp," Guimbail explained, the intellectual and the "brain worker" would either seek out opiates or be unwilling to let them go once they were introduced to them. Those not predisposed to intellectual excess, the logic went, would express a natural and healthy revulsion to opiates and their mental effects; since they would not enjoy the experience, they would not want to use the drugs any longer than medically necessary.[64]

Thus, even as doctors recognized that opiate addiction could have its origins in therapeutic practice, by the beginning of the twentieth century many believed that only patients with certain personality traits would succumb to opium or morphine habits. Abnormal character and addictive disorder, therefore, went hand in hand, and even though drug dependence had physiological underpinnings, it was still seen as a problem rooted as much in the psyche and the personality as in the body. Consequently, even with the recognition that many addicts had not necessarily sought out opiates for their psychotropic effects, it still took a certain type of person—hereditarily tainted, impulsive, or hyperintellectual—to fall victim to the drugs' charms. Just as alcoholism, according to many commentators, highlighted the gluttonous side of the drinker, so opiate addiction could stamp the character of the user as psychologically deficient, or greedy, or overly eager to procure the stimulation that opiates provided. The opiate addict was consequently seen as a victim of disease but also as someone who, due to character and disposition, was often deserving of his affliction.

Addiction as Metaphor

As Susan Sontag observes in *Illness as Metaphor*, disease is often more than just a medical understanding of pathological biological processes. It can also serve as a discursive "cheap shot" that marks and marginalizes the morally or socially suspect. Sontag highlights how cancer, for example, has historically been considered a sign of a repressed character, and tuberculosis a marker of an overly romantic disposition. In the late nineteenth and early twentieth centuries, the medicalization of many pathological and "deviant" behaviors, including substance abuse, was highly imbued with judgment of the character of those afflicted by psychiatric or behavioral illnesses. This was particularly the case as researchers began elaborating theories about substance abuse, because the perceived susceptibility to alcohol and opiate addiction corresponded to what researchers understood to be the subjective experience procured by each substance. Alcohol was considered a social substance that procured fleeting bodily and social pleasures, so alcoholism was a disease that tended to ensnare the gluttonous or irresponsible. Opiates were seen as substances that facilitated intellectual stimulation and a retreat from the outside world, so opiate addiction was a disease believed, in many cases, to strike at the hyperintellectual, the dreamer, or the snob. Addiction to each of these substances was both a disease and a marker of personality, one that could indicate physical and psychological disorder and a flawed moral makeup.[65]

The Imagined Community of Opium

By the end of the nineteenth century, the reputation of opium as an anti-social intoxicant had become ensconced in both the literary and medical discourses of Britain and France. Subjectively, the writings of De Quincey and studies on opium's psychoactive effects spoke of how the drug isolated users from those around them, both physically and emotionally. Objectively, the writings of medical researchers who elaborated theories of opiate addiction seemingly created a disease model that framed the disorder as a metaphorical expression of character flaws and reclusive tendencies.

Though considering it inherently anti-social, contemporaries did not believe habitual opiate use lacked social significance. To the contrary, according to nineteenth-century writers, opiate users formed a "society of dreamers," a group of individuals who were seemingly united by their antisocial tendencies. A fraternity of pariahs with common egotistical sensibilities, opiate users were seen as a group of individuals who were highly preoccupied with their own intellectual journeys but nonetheless shared common bonds of self-absorption, interiorization, and alienation. Thus opiates developed a reputation as an emblem and pastime for outcasts and recluses, individuals who were, paradoxically, united by their radical individualism and passion for personalized reverie. Even as the drug's inward thrust made users, as one commentator put it, "no longer belong to society," they were still bound by the "strong and jealous ties" of egotism that the substance embodied. "Opium," Farrère explained, was no mere pastime; it was "a fatherland" and a "religion." As a practice that united the self-centered, opiate use seemingly brought those with a common interest in this fundamentally selfish exercise into a subculture of their own. To be an opiate adept was not just to tune out of the society of ordinary men but to tune into a self-centered society of solipsists—individuals bound not by common blood, soil, or tradition but by an intellectual and philosophical disposition that led them to expel the outside world from consciousness and undertake psychotropic journeys.[66]

Opiate users thus came to compose what Baudelaire termed a "meditative nation lost within the heart of the busy nation," an "imagined community" of contemplative egoists who thought little of citizenship or solidarity but shared an unquenchable thirst for the imagined. Unconcerned with their neighbors and countrymen as defined by blood, soil, or spirit, these individuals seemingly answered the call of different flags: those of the self and the imagination instead of the Union Jack or the Tricolor. Defined by their desire to let their minds run free, the "meditative nation" of opiate users thus came to be seen as distinct from those who did not recreationally indulge in the substances. Opiates differed dramatically from the patriotic poison of alcohol in both their physical and psychological effects, and opiate habits were seen as vices of the Orient rather than as British or French traditions. Consequently, opium users came to constitute their own "nation" of self-absorbed dreamers within the broader community and discursively represented quintessential anti-citizens. Whereas the ideal citizen was connected to his cohort in a spirit of goodwill and bound by a sense of moral obligation to act for the good of the community, the opiate-using dreamer had no interest in social interaction or the well-being of those around him. What seemed to tie opiate users together, rather, was a desire to have no social ties whatsoever.[67]

Discursively marking users as both poor citizens and poor nationals in the late-nineteenth-century imagination, opiates represented a threat that encompassed more than the simple dangers of addiction and its attendant social problems. To take the drugs regularly was to disturb the balance necessary, for the ideal *Homo collectivus*, between individual and collective sensibilities. By so undermining social consciousness, the drugs became chemical embodiments of the qualities of poor citizenship and self-absorption run amok. Thus the problems posed by recreational opiate use were threats, not only to public health, but also to the essence of citizenship and the sentiments of national solidarity that came with it. Though the issues brought up by opiate habits were certainly medical and philosophical, at their roots, they were also social and political.[68]

CHAPTER TWO

Anti-narcotic Nationalism

The Feared Consequences of Recreational Opiate Use

> The whole nation will become one vast De Quincey—everyone will neglect his work.
>
> —*Macmillan's Magazine*

Opiate use was seen as a distinctive practice by the end of the nineteenth century, on two fronts—one spiritual, the other medical. Understandings of the psychoactive effects of opiates led many within both the medical and literary communities to consider the drugs agents of isolation that cut users off from the outside world, physically, psychologically, and spiritually. Unlike other chemical pastimes such as drinking alcohol, which was believed to enhance social existence, opiate use seemed to inhibit it. Furthermore, the chains of addiction that bound recreational users to opiates were widely considered to be stronger than the forces that tied the drinker to the bottle, and thus much more dangerous.

Opiates could have consequences for the health and personality of their users, but could also have a deeply political significance, in two major respects. First, given that the drugs could bring about the physical and mental ruin of those who indulged in them, they encapsulated one of the critical tensions of modern liberty: does the individual have a right to do what he wants with his own body, even if it endangers his health? Was drug use what John Stuart Mill deemed a "self-regarding" behavior that should not be regulated, or were its ravages significant enough that society should step in and protect individuals from their own narcotic appetites? Second, since opiates were believed to isolate users from those around them, making them both less aware of and less engaged with others, use of the substances brought up questions about citizenship and social inclusion. By facilitating a retreat from the social and toward the interior, opiates had the potential to alter the relationship between the individual and society, transforming citizens into isolated individualists who were both unwilling and unable to contribute to, or even participate in, social and political activity. Thus opiates became what anthropologist Alain Ehrenberg terms "zones of mediation between abstract categories," tangible substances that, because of the culturally and scientifically constructed meanings assigned to them, embodied tensions that were not just medical and philosophical but also fundamentally political.[1]

Different ways of thinking about society in Britain and France led to different ways of conceptualizing the nature of the opiate problem in the late nineteenth century. The different makeup of their opiate-using populations also shaped how Britain and France viewed their drug problem. Thus, by the beginning of the twentieth century, the two countries did not face a common drug *problem* but rather two different and nationally distinct drug *problems.*

British Anti-narcotic Nationalism

The State of the Opiate Habit in Late-Nineteenth-Century Britain

Some empirical evidence suggests that opiate use was becoming less problematic in late-nineteenth-century Britain. Though British opium imports increased dramatically, from 113,140 pounds in 1827 to 833,390 pounds in 1900, starting in the 1890s, a large amount of the imported opium was transformed into morphine and reexported, so it is difficult to ascertain how much went to the home market. Historian Virginia Berridge estimates that before the rise in pharmaceutical production, average estimated opium consumption dropped precipitously, from just below eleven pounds per thousand population per year in the mid-1870s to just over three pounds per thousand annually by the early 1890s, before rebounding later in the decade. Furthermore, statistics from treatment facilities reveal that it was relatively rare for individuals to enter treatment for opiate addiction in Britain during this time; an average of just two drug addicts found their way into treatment each year at the Dalrymple House at Rickmansworth, a facility that specialized in the treatment of alcoholics. Norman Kerr, in his 1889 study of substance use disorders, found that opiate habits were still relatively minor in Britain, concluding that for every opiate addict in England there were probably thirty in the United States. Increased awareness of opium's dangers, better labeling practices required by the 1868 Pharmacy Act, and increased use of other drugs in medical practice may have combined to decrease overall levels of domestic opium consumption.[2]

There is also evidence, however, that excessive opiate use was still common in late-nineteenth-century Britain. Many medical practitioners continued to dispense the drugs liberally, particularly in rural areas where medical knowledge and techniques lagged behind. Mortality statistics from the era suggest that heavy use may have increased among adults; though the prevalence of child opium overdoses decreased in the late nineteenth century, the rate of adult deaths from narcotics increased greatly between 1863 and 1890. In the 1880s, some doctors observed that nonmedical opiate use and opiate addiction were becoming more common. S. A. K. Strahan of the Northampton Asylum wrote in 1884 that opiate abuse was "a growing disease in this country." As Kerr elaborated, opium smoking, the consumption of pills containing opium, and morphine injections were becoming more prevalent among males, while laudanum use was increasing among females—facts that he claimed were borne out by overdose mortality statistics. Addiction specialists recognized that opiate abuse was probably more commonplace than they knew, since the habit was, by definition, a "secret one," and only a small proportion of those with serious drug problems ever admitted to their addictions or sought out treatment for them.[3]

Yet, even in situations where opiates did cause harm, they did not seem particularly menacing at this time. As historian Terry Parssinen found in his review of the British medical press of the late nineteenth and early twentieth centuries, addicts and the nature of their addictions—at least, those that were recognized and discussed—were not framed as major social problems by most doctors. Cases of opium addiction were rarely examined in detail, but morphine addiction was mentioned more often in the medical press, with fifty-one case studies profiled in British medical publications between 1870 and 1920. Most of the addicts discussed in these articles were well-to-do professionals or businessmen who did not seek out opiates for their psychotropic qualities, but started taking them in the course of medical treatment. Many of them began to experience physical and psychological troubles as a result of their addictions, but most were able to continue to work and lead happy home lives, often concealing their drug habits even from their spouses. Thus, even while addicted to opiates, these individuals continued to function as regular, contributing members of society.[4]

The cases cited in the medical literature do not necessarily capture the entire scope and character of the problem, however, since it was generally individuals of high socioeconomic standing who received treatment from the doctors who wrote research articles. Furthermore, given that morphine users tended to be from higher social strata and users of opium and opium tonics were more socioeconomically diverse, it is possible that overall opiate use was more widespread than the medical literature suggested—just not among those patients that doctors regularly saw. Even if opiate habits were more common than doctors knew, those that were recognized did not cause much concern in Victorian Britain, as it was still considered normal for households to have laudanum preparations in their medicine cabinets. Drug taking, as Berridge points out, was "never, at this stage, a way of life in itself."[5]

Nonetheless, anxieties about opiates—and smoked opium in particular—began to emerge in the late nineteenth century. British concerns did not focus on the current state of affairs or the actions of current users but rather on what opium had the potential to do in the future—both to individual Britons and to Britain as a whole—if indulgence in the drug became too widespread. Though informed by general scientific and cultural understandings of opium that were common throughout Europe, the British drug discourses that emerged at this time were articulated through and reflective of more general ways of British thinking about citizens, the nation, and the effects that opiates could have on both. Opiate use, while not yet seen as a major social or public health problem, was still regarded as potentially problematic, both politically and socially, within the framework of citizenship as it was understood in Victorian Britain.

Imagining Britain

From the eighteenth to the early twentieth century, British society was structured more loosely than its continental counterparts, and the main task of the sociopolitical order was to protect private citizens' individual interests. The British state focused on maximizing local autonomy and letting elites, generally at the local level, steer policy formation and implementation. The main manifestation of liberty, as Britons understood it, was commerce, which was believed to be both the engine of progress and the means of maximizing the public good. The job of government was to maximize economic opportunity by fostering trade and encouraging industry, making the state a utilitarian ally of the entrepreneur. In economic matters, this often meant adopting a laissez-faire approach, leaving industrialization and infrastructural development in private and local hands and adopting economic policies that fostered business and free trade. Even when the state was interventionist—engaging in colonial expansion, establishing an income tax, or passing social welfare legislation—this was done in the name of preserving and strengthening the economic order, expanding markets, protecting free trade, or neutralizing potential class tensions. As historians have shown, in a variety of arenas ranging from public health and social welfare to national defense and the creation of a telegraph system, when the British state intervened in private affairs, it was generally to enhance and expand individualism, a smoothly functioning market, and a prosperous economy.[6]

Cultural ideas of "Britishness" reflected social, political, and economic realities, as the nation came to be defined by its dedication to freedom and liberty and its existence as a "free acting, atomistic society." A good Briton was, above all, an individualist and a consumer, a self-sufficient citizen who was able to prosper in and contribute to the economy, adding to national wealth without becoming a burden to his countrymen. The key element that unified the nation was dedication to economic freedom, something that Britons believed made their nation different from the foreign powers that posed threats, real or perceived, to their freedom of trade. The bugaboo of French authoritarianism, as historian Linda Colley shows, was especially critical in the formation of British identity: the absolutist kings and Napoleon represented what Colley terms a "militarist, decadent, and unfree" threat to British liberty. "Once confronted with an obviously alien 'Them,'" she explains, Britons came to see themselves as a "reassuring or merely desperate 'Us.'"[7]

As the centuries of war with France drew to a close in the early 1800s, the lands of the East provided the next grounds for the construction of an oppositional "Other" that served as a yardstick against which British freedoms would be measured. Thanks in large part to stereotypes and excuses mobilized to justify the use of force in Asia, Britons came to see the "Orient," and China in particular, as irrational, despotic, backward, and unfree. Furthermore, the Chinese government's attempts to assert national sovereignty by placing limits on British commerce during the Opium Wars were seen as affronts to British liberty and prosperity, and as direct threats to what Britons held dear. Even though the Chinese represented little real threat to British autonomy, some Britons began to fear that the Chinese would rise up against Britain's military, mercantile, and commercial might and avenge their defeats. Such a Chinese "invasion" of Britain, however, would not involve direct armed confrontation; rather, it would be a gradual and subtle war of attrition, to be undertaken through a corrosion of British culture from within. These anxieties grew in the mid to late nineteenth century, as Chinese sailors and laborers immigrated into the dock districts of major cities such as London and Liverpool, and a Chinese subculture emerged in these areas. For the xenophobic alarmists who feared a slow and steady Chinese conquest of Britain, the increased Chinese presence—both numerically and culturally—seemed testament to the reality of what they considered a nascent Asian threat.[8]

Considered despotic, opposed to freedom of commerce, and hostile to British strength and prosperity, the Chinese became an ideological opposite, one that helped define British identity in the late nineteenth century. The Sino-phobic undercurrent in British culture would play a key role in the development of British drug discourses in this period, as the association of the Chinese with opium colored the nature of the threat that Britons feared the drug could pose to their national community.

Britain's Narcotic Nightmare

Within the context of British imaginings of the national community, opium had the potential to become a dual threat: it was believed to make users apathetic and lazy, traits that ran counter to the individualist and capitalist ethos that was thought to make Britain distinctive, and it was intimately linked to the Chinese, who represented a threat to the free and resourceful ways of the industrious Briton. Drug discourses emerging in the mid to late nineteenth century reflected both of these concerns, painting a picture of opium that made its habitual use seem not only distasteful but potentially dangerous for the nation and its citizens. These anxieties were articulated in two forums: in popular journalistic and fictional writings and in the work of pressure groups that sought to end the Indo-Chinese opium trade.

As the Chinese immigrant community became increasingly visible in Britain, some British commentators saw opium dens in Chinese neighborhoods as Oriental outposts stationed on the British homeland. In particular, the opium dens of London's East End became spaces that fascinated writers, places that were exotic but located near the heart of the nation. At first, the opium den was a subject of curiosity as much as one of alarm or reproach. In 1864, for example, a *Daily News* journalist told the story of his accidental discovery of a London opium den, but he was more intrigued by it than disgusted. The majority of the den's inhabitants, the writer found, were respectable, with a "good-tempered" and "comic" manner that was "far from unpleasing." The den the correspondent described was not a pure den of iniquity. It had representatives from all walks of life; businessmen, performers, clerks, and seamen peacefully coexisted within the space, each tending to their own affairs. Even the manager of the den, a Chinese man, was described as a respectable and seemingly industrious shop owner, who awoke early every morning to clean his establishment and make it welcoming for his patrons. The one potential drawback, the writer observed, was the effect that opium could have on some of the British users who frequented the den. Describing an Englishwoman who had become addicted to opium at the den, the writer wrote that "from long consorting with Chinamen," she had "acquired their habits" and become overly accustomed to their foreign ways. Thus, though the drug itself was not necessarily dangerous, it did have the potential to ensnare Britons who dabbled in it and to lead them to assimilate Chinese ways.[9]

As the 1860s progressed, attitudes toward Asians and the opium dens they frequented took a turn to the dramatic and alarmist. Charles Dickens, in particular, did not mince words in his description of Chinese immigrants and the dangers he believed they and their opium habits posed for Britain. In "Lazarus, Lotus Eating," a story published in 1866, Dickens recounted a fictional foray into London's East End. In place of the portrait of a respectable establishment as described by the *Daily News* just two years earlier, Dickens described the opium den as a "wretched hovel" that housed a racial grab-bag of exotic foreigners, including beggars, thieves, and a listless collection of individuals whose "sole bond" was a desire to "smoke themselves into pleasant satisfaction."

In Dickens's story, a Chinese man named Yahee orchestrates the whole operation. Yahee is far from a hard-working businessman: he is "always in a stupor," and having indulged in too much opium for too long, he seems more like a "hideous and long-forgotten mummy" than an industrious entrepreneur. Unlike the den proprietor described by the *Daily News*, Yahee is "indifferent" to his work and succeeds in his business only because of his secret method for preparing opium, which came to him when he entered a mysterious "cataleptic trance." Yahee's powers seem not only supernatural but expansive and predatory. Dickens describes one woman in the den, a former "London Lady" repeatedly lured into Yahee's establishment until she becomes a regular, who has become "Orientalized" by opium, picking up Chinese customs, language, and, most importantly, opium habits. Now working as a prostitute and peddler for Yahee, she has completely fallen from her standing in the British elite and become a puppet of the Chinese immigrant who gained her allegiance by feeding her opium.[10]

Dickens's "Lazarus" story was indicative of a shift in the way British writers portrayed opium dens and helped set the tone for subsequent descriptions. Whereas the opium den was merely a curious place in earlier depictions, the Dickensian den was a thoroughly threatening locale on many levels. For one, it reinforced the idea that Asians who indulged in opium were quintessentially un-British—lazy, indolent, and mystical rather than industrious, hardworking, and rational. Moreover, for Dickens, the opium den was not only a focus of filth and immorality but a hub of Oriental contagion and the gradual conquest of British culture; by describing the former "Lady" who becomes a prostitute, Dickens elaborated an image of opium as a subtle agent of Chinese "invasion" of British ways and conquest of British values. Transforming a genteel woman into a prostitute with the appetites, customs, and apparent "immorality" of an Asian, opium seemingly assimilated Britons into Chinese culture. Thus the drug had the potential both to intoxicate the British and to facilitate their slow and steady decline, by transforming Britons into what stereotype deemed to be their diametric opposite—the Chinese.[11]

Subsequent descriptions of opium dens built on the tropes of degradation and anti-Asian fears elaborated in Dickens's "Lazarus." Popular works such as Oscar Wilde's "The Picture of Dorian Gray," Arthur Conan Doyle's Sherlock Holmes adventure "The Man with the Twisted Lip," and the short story "A Night in an Opium Den" that appeared in *Strand Magazine* in 1891, all featured descriptions of opium dens that reflected the image portrayed by Dickens. Authors after Dickens shrouded their opium dens in mystery, generally locating them in seedy neighborhoods, behind docks, or at the ends of ominous, dark staircases. The entrances to the dens were also eerie, described as "caves" hidden behind trapdoors that required a secret knock to secure entry. The insides of these dens did not make them any more comforting, as poor lighting, filth, and bizarre decorations, such as disturbingly blasphemous pictures of the crucifixion, bore witness to their state of physical and moral decay.[12]

The decor of these fictional opium dens, however, was delightful when compared with the clientele. Some were described as foreign and savage—Malays crouched in corners or half-naked Asians lying in "strange fantastic poses." The denizens of the opium dens had few scruples, as they were willing to steal a man's hat, boots, or umbrella while he slept off his opium-induced trance. Human beings under opium's spell morphed into dangerous animals, becoming "coiled up" and "furtively watching" like "pythons in a serpent house watch the visitors who come to tap at the glass at their cages." Others seemed to become lifeless altogether, reduced to "grotesque things" with gaping mouths and dull eyes that made them seem more sculpture than human. Opium, in these descriptions, had the power to cause users not just to decay but to degenerate, as it transformed the once human (albeit "inferior") Asian into an animal-like or, even worse, lifeless creature.[13]

In these depictions, while seemingly destroying the Asians who frequented the dens, opium could also transform the Briton into an "Oriental." Conan Doyle, for example, lamented how opium had turned a once-distinguished theological scholar into a "wreck" with distinctively Asian features, such as a "yellow, pasty face, drooping lids, and pin-point pupils." The author of "A Night in the Opium Den" took the metaphor of "yellowness" to another level, writing that when under the influence of the drug, the "whites" of one's eyes would more appropriately be dubbed "the yellows."[14] Yet the equation of opium with decay among British users was not one of simple transference (to be Asian was to be un-British; to smoke opium made one Asian; ergo, opium was un-British). In "The Man with the Twisted Lip," in particular, opium served as a plot device to illustrate more than a mere "Orientalizaton" of Caucasian characters who dabbled in the drug.

"The Man with the Twisted Lip" revolves around the double life led by Neville St. Clair, a respectable suburbanite who, unbeknownst to his family and friends, works as a beggar in the heart of London. St. Clair begins panhandling as part of his job, working as a journalist writing a story about homelessness. For the story, he decides to try begging for a day to get firsthand experience on the subject. Soon, he realizes that he can make more money panhandling in one day than he can earn in a week as a journalist, and he begins begging full-time. Eventually, his begging earns him enough to marry and buy a house in the country. To maintain his double life, St. Clair rents a room in an opium den, which he uses each day as a changing room to transform himself from respectable businessman to homeless beggar. In the den, he removes his collar and tie and puts on an elaborate mask of makeup and plaster to give him the appearance of a transient. Then, in the evening, he returns to this room, removes his disguise, and dons his business attire before returning home. Thus the opium den serves as a locus of transformation for St. Clair, a place where he can completely change his identity in secret while maintaining its duality.[15]

Beyond serving as a private space where St. Clair can disguise himself, the den is also a place where he undergoes a metamorphosis that was significant in British culture—from businessman to beggar. While he lives much of his life as a man who is the consummate self-sufficient and enterprising Briton, St. Clair is actually the opposite: an individual who relies on charity to survive. The transformation that St. Clair undergoes in the opium den, therefore, is not simply between middle-class comfort and indigence; it is a constant shift between being a self-sufficient member of the community and a dependent, parasitic tramp. In this scenario, the opium den serves as the space where the prototypical self-reliant man becomes a charity case; or, in British understandings of good citizenship, it is a place where good citizens can become bad citizens.

While literature reflected and reinforced opium's reputation as a substance of the Oriental and the indolent, some nonfictional works also helped to shape the drug's reputation in Britain. The principal source of these writings was the anti-opium lobby, a religiously inspired movement that argued that the British Indo-Chinese opium trade should end for moral reasons. Unease about the ethics of the opium trade had been prevalent throughout the nineteenth century, but in the 1870s the anti-opium movement became a better organized and politically powerful force, with creation of the Society for the Suppression of the Opium Trade (SSOT). The SSOT gained widespread support, not only from its Quaker base, but from Methodists, Baptists, Presbyterians, Unitarians, and other dissenting churches. Spreading their message through public meetings and regular publications, the "anti-opiumists" of the SSOT sought to sway British public opinion—and eventually British policy—against what they considered to be the pernicious eastern trade in opium. In the process, the SSOT also facilitated a good amount of reflection and debate concerning the nature of opium itself.[16]

The SSOT made its case against the opium trade along two interrelated axes, one concerning the drug, the other concerning Britain's colonial mission. "Missionaries attack the trade," SSOT leader and ex-missionary Storrs Turner explained, "because their intercourse with natives convinces them that it does them grave harm." In India, where the poppies were grown, as well as in China and Burma where most opium was eventually sold, the SSOT maintained that the drug's effects were detrimental to the well-being of both individual consumers and society as a whole. This was largely due to the toll that opium took on long-term users, which in the descriptions provided by missionaries mirrored the nightmarish effects that doctors hypothesized excessive opium use could cause. According to the missionaries, the opium habit engendered neglect of one's duties to work and family, facilitated indolence, and quickly led down a road toward unemployment and a painful alternation between destitution and opium-induced intoxication. The end result of most recreational opium use, warned the SSOT, was a tragic one, as the drug would make users vulnerable to disease, emaciate them, and facilitate a lifestyle that landed them either in the hospital or in jail. Furthermore, the opium habit was seen as so widespread that it was threatening to consume entire swaths of the Asian population and instigate an epidemic of madness, stunted population growth, and even extinction. Not only did the drug damage the individuals who consumed it, the SSOT believed, but it also compromised the religious and ethical justifications for Britain's imperial enterprise by poisoning, rather than enlightening, Asian peoples.[17]

By engaging in such a morally reprehensible enterprise, the SSOT warned, Britain was setting itself up for failure in Asia and for a vengeful justice that would eventually hit home. "If nations permit great national sins," the Lord Mayor of London, a supporter of the SSOT, sternly warned, "they will be punished for them. There is a time of retributions for nations." By fomenting distrust of the British among the Chinese, in particular, the opium trade had the potential to make Britain a sworn enemy of the large and potentially powerful empires of the East. Given that China might one day evolve into a global force, SSOT sympathizers believed, it was important for the British to work for, rather than against, the interests of the Chinese. If not, warned SSOT supporter Cardinal Manning, a "scourge" could lie in wait on Asian soil, "preparing for us if we alienate the Oriental races." By highlighting the dangers that could be developing in Asia, the SSOT mobilized a broader fear of the Chinese and a potential Chinese "invasion." Opium, in this scenario, was not just a potential means for a Chinese attack against Britain—as it was in fictional accounts—but also its cause. Blending moral suasion with fear of vengeance, the SSOT made an argument that could appeal to both the decency and the oppositionally defined national identity of Britain. Stopping the opium trade, according to this line of thinking, was not only right but imperative.[18]

Through its propaganda, the SSOT spread ideas about the opium trade throughout Britain and also contributed to the general discourses around opium. Using China as an example of what could happen to a society if the opium habit were allowed to spread, the SSOT provided a nightmarish vision of what an opium-drenched Britain could look like: degenerate, depraved, and in decline. Elaborating on the discursive edifice that had linked opium with the foreign and the indolent in fiction, the SSOT's propaganda helped entrench in the British mind the equation of opium with the Orient and the potential demise of the nation. Together, these narcotic nightmares integrated two critical strands of British thought about the national community and why opium posed a threat to it. With their links to laziness and dependence, opium and opium dens became the chemical and physical antitheses of the British values of industriousness and independence; with its links to the Orient and decay, as elaborated in both fictional accounts and SSOT propaganda, opium was integrated into broader anxieties about China, the demonic "Other" that conceptually grounded Britons' understandings of themselves as a nation.

Opium on Trial: The Royal Commission on Opium

By the 1880s, the SSOT had secured the support of many members of Parliament, who introduced (but failed to pass) resolutions calling for the abolition of the opium trade in 1875, 1880, 1883, 1886, and 1890. Yet as the SSOT's influence over some in Parliament grew, so did a movement against the society, which argued that the Indo-Chinese opium trade was acceptable and that the SSOT was at best misguided and at worst intentionally deceiving the public. Predominantly a collection of colonial doctors, administrators, and businessmen, this group did not form an organized lobby or movement like the SSOT, though many of them did publish tracts and pressure politicians to maintain the status quo for India's opium business. Their main tactic was to depict the SSOT as a group of religious and imperialist fanatics who were either ignorant of the effects of opium on consumers or culturally insensitive to the fact that the opium habit was an accepted pastime in the East. The drug, they argued, was essential as a medicine in China and India, and for many people it was often the only reliable defense against disease. Apologists for the opium trade also maintained that the drug was the most natural psychoactive substance for Asian peoples, one that fit their lifestyle and traditions—much as alcohol did for Britons at home. "In the same way as the honest sons of Britain would express their unqualified disapprobation and indignantly resent any attempt to deprive them of beer," one unabashed critic of the anti-opium lobby, explained, "so would the inhabitants of India, and of China, resent, if not resist, any interference with their use of opium."[19]

According to those who wanted to maintain the status quo, opium not only was cherished in Asia but it was not nearly as damaging as the SSOT propaganda implied. In both India and China, they argued, opium users were robust, physically healthy, and of a "most chivalrous and romantic temperament," rather than the decrepit individuals who peopled the anti-opiumists' horror stories and fictional opium dens. Colonial doctors, in fact, often saw Asians who had taken the drug regularly for up to forty or fifty years and showed few ill effects. In cases where opium users did suffer illness or social problems, critics of the anti-opiumists argued, this was usually because of something other than opium itself—a brain injury, a chronic disease, or a faulty character. For these individuals, the opium pipe "was merely the last straw laid on their already enervated and overstrained backs." To bring this point home, many writers compared opium's effects to those of alcohol, concluding that opium was not nearly as destructive as drink. Not only did alcohol inflict more physical damage on the body, they claimed, but it also caused more social disruption than the opium habit. Whereas a man under the influence of opium would peacefully doze in a trance-like state during his intoxication, the drinker was apt to return home from a drunken debauch to physically abuse his family, wreak destruction on his own property, or get into a fight. Thus when compared with alcohol, they concluded, "opium is the lesser evil of the two." Some writers even went as far as to challenge SSOT leaders to simply try the drug themselves if they wanted to learn about its effects firsthand. The SSOT's claims, they charged, were but a series of misrepresentations, half-truths, and distortions that culminated in "one *concrete* . . . *fascinating, defamatory lie*," which, though "sentimental and pious," was but a "sham, a mockery," and "a delusion." The dreaded effects of opium described by the SSOT were based on "sensational exhortation and sentimental harangue," built largely on religious fanaticism rather than fact or reason.[20]

Parliament called for a royal commission to study the opium question in June 1893. Accepting it as fact that the Chinese would still use opium even if Britain ceased production in India, the Royal Commission on Opium focused on two sets of questions: one concerning India, the other concerning opium. On the first, the administrative issues in India, the commission set out to determine whether the growth and manufacture of opium in India should be prohibited (except for medical purposes), what effects such a prohibition would have on the Indian economy, and whether the current opium regime should be modified so that it could maintain revenues while diminishing opium consumption. The second set of concerns focused on the drug itself and its effect on the moral and physical well-being of individuals who consumed it. Holding hearings in London, India, and Burma, and interviewing 723 witnesses who answered a total of twenty-eight thousand questions that filled nearly twenty-five hundred pages, the Royal Commission gathered a massive amount of evidence that would inform how the British viewed not only the questions relating to opium and the finances of India but also the nature of the drug itself.[21]

The most highly represented profession among the commission's witnesses was medical practitioners, who provided 146 testimonies; government officials, military officers, landowners, missionaries, teachers, and representatives of private associations also contributed heavily to the body of evidence. The opponents of the anti-opiumists did not have any official representatives speak before the Royal Commission, but many witnesses—particularly those who worked for the Indian government and did not want to see its finances disrupted—supported their cause. On the other side, the SSOT was sure to make its voice heard, as 152 of its members provided testimony—meaning that more than twenty percent of the evidence heard by the commission had the anti-opiumists' stamp of approval. While the SSOT was more effective in organizing witnesses, however, the composition of the commission itself was evenly split between anti-opiumists and their critics.[22]

The testimonies in favor of the anti-opiumist cause, though numerous, were not particularly convincing to the Royal Commission. The commission's assignment was to study opium's effects on India, but many of the anti-opiumist witnesses focused their testimonies on China, where most Indian opium was consumed, and on the effects of the drug there. Moreover, anti-opiumist arguments proved to be, as critics had said they were, more "sentimental harangue" than reasoned argument. The majority of the anti-opiumist testimony was based less on science than on emotional appeal and hyperbole, describing opium as a supernatural power that acted as a "vampire" on the body and soul. Their evidence also tended to rely on the testimony of missionaries rather than on scientific literature, and their points were rarely backed up with medical proof. This evoked a good deal of skepticism from members of the commission. "How can you explain," one commissioner challenged during an anti-opiumist's statement, "that many opium smokers are merchants, clerks, tradesmen, artisans, and laborers who are . . . very shrewd, industrious, and successful . . . [and] notorious among Asiatics for honesty?" When confronted with such questions, the anti-opiumists could offer little in the way of a convincing counterargument or comeback. Consequently, many on the commission simply dismissed the anti-opiumists' arguments as mere "terrible tales" rather than hard evidence.[23]

While the anti-opiumists' argument fell flat, their opposition's case proved sturdier under scrutiny. The anti-opiumists' testimony came mostly from missionaries and was tainted by exaggeration, but their critics tended to be competent and sober in their presentations. Since almost all of the opponents of the SSOT had lived in India or China for an extensive period, they were able to speak with authority on the lifestyles of Asian peoples, and since they were doctors and colonial administrators, they were armed with the scientific and statistical arsenal needed to back up their assertions. They acknowledged that excessive opium use could cause serious problems, but maintained that addiction was more the exception among Asian opium users than the rule. Although opium could be abused, witnesses argued, most users indulged in the drug in moderation, and the practice had as little negative effect as the drinking of wine in France or Italy. "It would be impossible to recognize a moderate opium-eater," a physician stationed in Calcutta explained as he attempted to describe the consequences of long-term use. "You might as well expect to recognize whether a man drinks tea or coffee." Far from being damaged by opium, most users remained healthy. In addition, witnesses emphasized that the opium habit did not inevitably lead to excess, as the vast majority of users never increased their dosage beyond a safe amount. The argument that the drug shortened life took a devastating blow when a fifty-eight-year old man who had been smoking opium since the age of nineteen provided eloquent testimony, and the argument equating opium with indolence was weakened when a member of the Indian Board of Revenue testified that even the heaviest opium users were just as fit for work as everyone else. Statistics presented by officials working for the Indian government also seemed to debunk the theory that opium caused insanity. In the asylums of Lower Bengal, for example, opium was identified as the trigger for madness in just eight of more than two thousand patients.[24]

To bring their point home, opponents of the SSOT argued that the consequences of opium use were not as damaging as those of drinking alcohol. "It never produces any disease, excepting in the very last stages of an opium-eater," colonial surgeons explained to the Royal Commission. Some highlighted the differences between the physiological consequences of drinking and opium smoking, others claimed that opium was less addictive, and yet others emphasized that opium did not cause the same sorts of unruly behavior as alcohol did. Though "the opium dens draw together a certain number of useless, worthless, idle fellows," John Lambert, Deputy Commissioner of Police in Calcutta, conceded, "I should not look there for any person whom I considered dangerous or habitually criminal." Statistics presented to the commission backed up this argument. Over the course of five years in India's Lower Provinces, opium was not believed to have facilitated any of the crimes of the 302,000 prisoners who were incarcerated.[25]

The anti-opium arguments, concluded many witnesses before the Royal Commission, were based on conjectures, exaggerations, mistruths, or some combination of the three. One Canton businessman, for example, argued that the entire anti-opiumist case was based on "generalities" that had little to do with the realities in Asia. Another witness agreed, saying that the anti-opiumist propaganda was patently ridiculous: "I do not believe you could get one intelligent unprejudiced Englishman who had lived for six months in Shanghai or any other part of China" who would find anti-opiumist argumentation to be anything but "a terrible perversion of the truth." As a British diplomat serving in Asia summed up before the commission, the SSOT's propaganda was as "disgracefully unjust" as it was "utterly false."[26]

When the Royal Commission on Opium issued its final report in 1895, the defeat of the anti-opiumist argument was clear. The commission wrote that both sides tended to overstate their cases during the inquiry, but it singled out the SSOT in particular, noting that the anti-opiumists had presented falsified evidence and petitions with fake signatures to make their medical arguments more convincing. Moreover, that non-Protestant missionaries, such as those from Roman Catholic and Syrian churches, did not agree that opium was such a problem also put the reliability of SSOT testimony in doubt. While acknowledging the sincerity of the anti-opiumists, the commission found that they tended to have "inadequate experience" and that most of their objections did not warrant consideration. The testimonies of physicians, however, were taken seriously. Based on their scientific evidence, the commission concluded that opium was, above all, "the common domestic medicine of the people" in Asia, an invaluable cure for disease and discomfort among those who did not have access to certified or trained doctors. Moreover, the commission found that the "preponderance" of the medical evidence supported the view that opium use was moderate and had "no evident ill effects" on the vast majority of users. Having seen many apparently healthy individuals who had been using opium regularly for an extended period, the commission found that the descriptions of what extended use could do, as told in anti-opiumist testimonies, did not add up. Furthermore, evidence showed that the drug did not tend to precipitate crime, insanity, or absenteeism and that its adverse social effects were not particularly pronounced. Even if it had found prohibition of nonmedical opium use to be advisable, the commission maintained, the obstacles in creating and enforcing such restrictions would be "insurmountable." Thus, in its conclusions, the commission decided there was no need to prohibit the cultivation or trade of nonmedical opiates in India, or to limit their exportation.[27]

The Royal Commission was one of the first forums in which the nature of opium and the consequences of its abuse were openly debated from both scientific and moral standpoints. The group's work, therefore, represented not just an inquiry into the specifics of the Indo-Chinese opium trade but, to an extent, a trial of opium itself. Was it indeed the horrible substance of foreignness, filth, and indolence that was depicted in popular fiction and SSOT propaganda? The Royal Commission answered this question with a resounding no, providing seemingly objective and convincing proof that the drug's reputation as a dangerous substance was generally undeserved. On this count, opium was largely innocent. The commission's findings made clear that opium was not necessarily incompatible with an industrious and self-sufficient lifestyle and was not always an indicator or instigator of stereotypical Asian laziness, backwardness, or decay. The drug was not necessarily as much of a threat to the British way of life as it was framed to be by the SSOT and other opponents of the opium trade.[28]

The conclusions of the Royal Commission's exploration into the nature of opium and addiction, when placed in a context where British domestic opiate abuse was still rare or rarely recognized, helped reframe the threats represented by opium use as relatively minor at the start of the twentieth century. In the early 1900s, the Royal Commission's conclusions on the effects of opium would be affirmed on the domestic front when investigations by local authorities into opium dens concluded that use of the drug, even in Chinese neighborhoods, was in no way inimical to productivity or self-sufficiency. Nonetheless, some of the nineteenth-century concerns about opiates would endure into the twentieth century. In 1909, the London County Council passed a by-law prohibiting opium smoking in licensed seamen's boarding houses where Chinese immigrants lived, and writings linking opium with Oriental decay and debauchery would reemerge after World War I. Yet, overall, the concerns about opiate use and its potential consequences remained relatively muted, especially when compared with the more dramatic drug discourses emerging on the other side of the English Channel.[29]

French Anti-narcotic Nationalism

The State of the Opiate Habit in Late-Nineteenth-Century France

In late-nineteenth-century France, unlike in Britain, contemporaries believed that opiate use and addiction were on the rise. Commentators had observed opiate use and addiction in rural and industrial areas of France earlier in the century, but the price of opium preparations was prohibitively high when compared with alcohol. Furthermore, a series of poisonous substances laws enacted in the 1840s restricted the right to sell opiates to pharmacists and individuals holding government licenses, thus placing another barrier between the drugs and most of the population. Consequently, opium habits remained relatively rare in France through the first half of the nineteenth century. But when access to medical care improved in the mid to late 1800s, so did access to the opiates commonly used in medicine, and overall levels of opiate consumption increased. The central pharmacy for Parisian hospitals, for example, used just over a quarter of a kilogram of morphine in 1855; this grew to more than 10 kilograms in 1875, and 750 kilograms by the early 1890s.[30]

Nonetheless, compared with Britain, overall levels of opiate use were still relatively small. On the basis of government statistics on opium imports and exports in 1907, in conjunction with population data from the era, it can be concluded that the British were still consuming about 4.3 times as much opium per capita as the French in the first decade of the twentieth century. But even though the overall amount of opiate consumption may have been less in France, opiate habits were much more visible to the French doctors who wrote up medical studies of addiction than they were to their British counterparts. Writing in 1889, Pichon analyzed 110 cases of morphine addiction that he had seen; in 1913, another doctor claimed he had cured 45 morphine addicts over the previous few years. These numbers are striking when compared with the mere 51 case studies in Parssinen's review of the entire British medical press over a fifty-year period. Furthermore, unlike British medical researchers, the French tended to used methods of statistical extrapolation to come up with extraordinary and alarming estimates of the prevalence of the drug habits. In Paris alone, one doctor estimated that in 1883 there were forty thousand morphine addicts; a few years later, another placed that number at sixty thousand. As historian Jean-Jacques Yvorel shows in his study of late-nineteenth-century medical literature, these estimates were absurd and based largely on faulty mathematical assumptions and small population samples. Nonetheless, these statistics, particularly the claim that there were sixty thousand morphine addicts living in Paris, became accepted as fact in many contemporary discussions of drug addiction.[31]

Beyond the perceived quantitative difference between French and British estimates of the scope and scale of opiate use, medical and epidemiological studies framed it as qualitatively different in France. Throughout French discussions of addiction in the late nineteenth and early twentieth centuries, a common concern voiced by medical researchers was the breadth of the population that appeared to be succumbing to the temptations of opiates. Whereas the British addicts described in medical studies tended to be respectable and of the middle classes, French addicts ran the gamut from highly respected doctors and social elites down the socioeconomic scale to artists, manual laborers, and habitués of the underworld. As Pichon observed in 1893, morphine addiction in France had been "barely heard of a dozen years ago," but it had spread rapidly through an "invasion" of "*all* social classes." In his 1889 sample of 110 addicts, Pichon found many middle-class businessmen and doctors addicted to morphine, but also eighteen laborers, thirteen demimondaines, two farmers, two housekeepers, and a nun. Doctors were not alone in observing the spread of the morphine habit, as semi-fictional works such as Maurice Talmeyr's *Les possédés de la morphine*, published in 1892, warned of its spread within and across social classes, trickling down from doctors, well-to-do housewives, and socialites to manual laborers and teenage street vendors.[32]

In addition to the expansion of morphine use, French commentators also believed that the character of morphine habits was changing in the late nineteenth century. Though they still acknowledged that most cases of morphine addiction were iatrogenic, French doctors became increasingly cognizant of addicts who had no medical reason to take morphine but began injecting it just to experience its pleasant psychoactive effects. The increasing visibility of recreational morphine use among elites in French nightlife, beginning in the 1880s, elicited concern from both medical and journalistic commentators, fueling fears that the morphine habit was a problem no longer limited to the unfortunate victims of disease and was becoming a fashionable vice.

Moreover, beginning in the 1890s, investigations by pharmacists revealed that some individuals who were using morphine recreationally did not secure the drug through regular medical channels, as was required under the poisonous substances laws; instead, they purchased morphine directly from wholesalers, or the drug was smuggled into the country on trains or by mail, thus avoiding any medical controls. Consequently, the use of morphine without the counsel of medical professionals—and thus opiate addiction—could have been spreading much more quickly than doctors and pharmacists realized. The rise in court cases in which defendants claimed that they stole to feed their morphine habits or committed crimes under the influence of the drug suggested that addiction was causing both medical and social problems.[33]

Even more worrying was that morphine was not the only form of opiate use that was spreading in the late nineteenth century. Opium smoking, especially in dens, was becoming a more prominent practice among French soldiers and functionaries stationed in Indochina, where the colonial government grew and sold the drug liberally. Though strongly discouraged from smoking opium, many sailors and colonial soldiers stationed in Indochina picked up the habit and brought it home with them. Servicemen smuggled in the drug when they returned to France, and crews aboard both military and commercial ships began to do the same. Soon, an underground opium network developed in major port cities such as Brest and Toulon, as demi-mondaines and Chinese knick-knack shops readily dealt the drug to soldiers, elites, and others who would set up small, private opium dens. In time, distribution networks and dens spread to Lyon, Paris, and other areas of the country. According to some critics, the emergence of the opium habit—an Asian vice—in France was a natural but unfortunate side effect of France's colonial mission in the Far East. As military physician Dr. Lefèvre lamented, "we have given the Indochinese our taxes and our alcohol; in exchange, they give us a taste for opium." By the turn of the century, critics such as Georges Claretié of *Le Figaro* would voice concern that the opium habit, like morphine, was beginning to "invade" France. As commentators began to take notice, they, like the doctors who gave very high numbers in their assessments of morphine addiction, began to sound the alarm by printing eye-catching estimates. In 1901, they suggested that there were fifty-six opium dens in Paris alone. A decade later, a more startling but highly implausible figure—twelve hundred Parisian opium dens—became the common number bandied about in discussions of the opium habit in France.[34]

Thus, in spite of statistics suggesting that opiate consumption was still more common in Britain, the apparent growth of the habit in France, its reach across the socioeconomic spectrum, and the more prevalent drug sub-cultures made French opiate habits seem much more menacing. Further-more, many in France came to see opiate habits as "contagious," claiming that as addicts boasted of opiates' pleasures, they would "proselytize" and recruit previously healthy individuals into the ranks of the addicted. By 1907, some French commentators assumed that opiate habits were, as journalist Delphi Fabrice claimed, "more contagious than leprosy on a ship." Thus, whereas in Britain, opiate use was only observed among a handful of iatrogenic addicts and immigrants and was considered well-contained, in France it was becom-ing increasingly prominent and garnered the attention of many in both the medical and mainstream press.[35]

The divergence between the perceived nature of the danger posed by opi-ates in Britain and France was not only a matter of demographics. The French believed that their opiate habits were more common, widespread, and expan-sive than in Britain, but they also had different fears of what the prolifera-tion of opiates could do if allowed to continue unabated. The opiate menace was particularly threatening for the French because of what they believed the drugs could do to the health of the nation—physically, ideologically, and politically—as it was understood in fin-de-siècle France.

Imagining France

Unlike the British, who envisioned the ideal relationship between the body politic and the individual as relatively loose, the French saw the collectivity of the nation as a more active player in the cultivation of individual liberty and prosperity. As far back as the days of the absolutist kings, both the French state and society were centralized, with Versailles serving as the undisputed center of cultural, political, and economic life. Although the French Revo-lution overturned the monarchy, it did not displace the important role of the central government in the lives of citizens. The revolutionaries of 1789 continued to venerate the state, considering it an entity that would not only govern but also play a key role in daily life; the main difference was that in the revolutionary order, citizens gained agency over the political process, and their active engagement in politics became essential for the success of the na-tional community. At the height of the Revolution, it became incumbent on citizens not just to obey the revolutionary order but to become active partici-pants in it, making the political a true extension of the personal. To truly be a member of the nation, the logic went, was not necessarily a matter of birth or geography but one of civic engagement with, and affection for, the social whole. Actions and demonstrations of devotion, therefore, were the keys to revolutionary citizenship.[36]

Though inclusive, citizenship was also demanding, requiring what sociologist Rogers Brubaker terms a "disciplined and exclusive political commitment to the fortunes of the state." Those who privileged private or sectional interests over those of the national community were considered suspect, and revolutionary violence was designed, in part, to eliminate such potential instigators of disunity. As historian Judith Surkis explains, within the revolutionary framework, citizenship was "not simply a political right, but also a normative construction."[37]

When stability was restored with the rise of Napoleon and subsequent regimes, centralization remained a unifying theme of both governance and cultural life. Monarchs, republicans, and emperors alike continued to focus on preventing sectional or selfish interests from supplanting the leading role of the state and society in the lives of citizens. In the Third Republic that emerged in 1870, political leaders worked to mobilize the revolutionary legacy, but in a tamer form, and to forge a nation that would be simultaneously democratic and more united—administratively, politically, and culturally—than its predecessors. The founders of the young Republic held that for a citizen to be a truly active contributor to the common weal, and not just a conformist, he needed a degree of autonomy that would allow him to contribute substantively to public life. The relationship between the individual and the social whole in the Third Republic was thus imagined to be a symbiotic one, with the state depending on the initiatives and efforts of citizens to provide expertise and leadership, and the people relying on the state's ability to protect citizens' rights and provide for them in times of need. A good French citizen, therefore, was to be both free within yet in solidarity with society, both hardworking and independent yet nationalist and civic-minded. The view of nationhood as a collectivist undertaking was further elaborated by Prime Minister Léon Bourgeois in 1895 with his theory of social solidarism, which called for the uniting of private and public interests into a cohesive and overarching whole. Citizens, for Bourgeois and the republican establishment, needed not just to support the state but to contribute to the common good; citizenship had its benefits, but it also came with duties. The strength of the nation, consequently, was only as great as the civic-mindedness of its components parts—the citizenry. Devotion to the national whole thus became more than just a matter of morality in the political universe of the Third Republic; it touched on the vitality and survival of the nation itself.[38]

Within this ideological context, the French state sought to fulfill what historian James Lehning terms a "dual project" during the early years of the Third Republic. A republican system required the establishment of democratic institutions, and it also needed to create citizens who would actively participate in and give strength to the political order. Consequently, the state worked to encourage citizens to embrace both the rights and the obligations of citizenship, turning all denizens of France "into Frenchmen" by putting the republican and transformative agenda of the Revolution into action. The main way to accomplish this was through state action, particularly in education, which was built around a curriculum designed to inculcate devotion to the well-being of the nation. Teaching children that it was their duty to work, learn, pay taxes, vote, serve in the military, and if necessary sacrifice their lives for the national defense, French schools served as institutions that sought to create a responsible and engaged citizenry. The leaders of the Third Republic also worked to minimize the influence of potentially divisive forces that could compromise loyalty to the social whole. With the passage of legislation designed to limit the influence of religious institutions in public life, with immigration policies that encouraged assimilation, and with a more centralized economy designed to advance the common good, the leaders of the Third Republic worked to ensure that no forces of a "selfish, particularistic, or antinational" character would compromise citizens' devotion to the collectivity of the nation. [39]

Imagining France as a thoroughly civic and republican nation, the Third Republic defined itself both positively and negatively—as a nation of engaged and responsible citizens and one that, by extension, could ill afford to tolerate those who did not buy into the assimilationist, communalist, and solidaristic ties that bound together the national community. Opiates, with their ability to cut users off from their cohort and immerse them in the psychotropic wonders of their own minds, held the potential to undermine these norms and expectations of French republican citizenship. This political aspect of opiates and their dangers became manifest in writings about the drugs in the late nineteenth and early twentieth centuries.[40]

France's Narcotic Nightmare

Just as in Britain, where anxieties about opium reflected larger national concerns about individualism and the Chinese, French drug discourses were colored by how the nation and its potential downfall were imagined. Fears about opiates were largely influenced by concern about their possible effects on the solidaristic bonds that both constituted and supported the strength of the nation. The arena in which these anxieties became most clear was the military, which in the late nineteenth century served as both the metaphorical expression and the physical protector of the nation. French fears of opiate use among the armed forces were more than simple manifestations of ideological anxieties, as the opium habit was not uncommon within the military ranks. But in a series of fictional works describing the moral effects of opiates on the military, authors not only warned of the prevalence of opiate habits within the armed forces but also contributed to understandings of what the ideological consequences of drug use could be.

Generally, the morphine habit was less intimately linked with worries about military decline than was the practice of opium smoking. In fictional works such as Marcel-Jacques Mallat de Bassilan's 1885 *La comtesse morphine*, for example, the drug served as an instigator of general decadence among French elites. Yet in affecting France's upper classes, morphine also had the potential to affect the behavior of military men, who sometimes floated in and out of high society. One work in which this danger was particularly manifest was Jean-Louis Dubut de Laforest's *Morphine: un roman contemporain*, published in 1891. *Morphine* told the story of French count, military captain, and morphine addict Raymond de Pontaillac. Originally turning to morphine after sustaining an injury in a duel, Pontaillac gradually becomes addicted to the drug. As related over the course of the book, morphine alters Pontaillac's personality, making him less engaged with what Dubut de Laforest terms his "exterior self" and more self-consciously "alone" in the world. His morphine-induced dreams and visions become increasingly warped and megalomaniacal over time, as he begins seeing images of voluptuous women when trying to study military maps, engages in strange Symbolist mental exercises, and devises a bizarre plan to create a new human species that can fly. Throughout these morphine-induced mental adventures, however, Pontaillac never becomes insane. "Not entirely split like neurotics," Pontaillac remains "in full possession of the self," only now fixated on the psychotropic wonders within his own mind instead of the real world. Though his thoughts and actions bear some marks of madness, Pontaillac is a victim not of mental illness as much as of an excessive pride that drives him to delusions. It is because of morphine that Pontaillac "rediscovered consciousness of his self" and loses touch with the world around him. Thus he is not insane, Dubut de la Forest concludes, but more of a "sinister and modern Don Quixote," a man who is blind to his own hubris yet remains driven by excessive pretension, imagination, and egoism. As his habit escalates, however, morphine starts to take a toll on Pontaillac's physical and mental health, and he is forced to quit the army and enter treatment.[41]

The story of Pontaillac's demise, though alarming, was not nearly as frightening as the tales of opium smoking in the naval and colonial forces that emerged in late-nineteenth- and early-twentieth-century writings. Fictional works recounting the experiences of military men who became entangled in the opium habit, including Pierre Custot's *Midship* and collections of vignettes such as Claude Farrère's *Black Opium* and Jules Boissière's *Fumeurs d'opium: comédiens ambulants*, voiced and helped contribute to French understandings of the opium habit and the dangers it could pose.[42]

In many of these stories, the tragic heroes are men predisposed to fall into the opium habit because of an overly contemplative mindset and a proclivity for reverie. The addicted protagonist of *Midship*, a young naval officer named Albert Douvesne, has a sensitive character that makes him susceptible to the allures of opium even before he tastes the drug. From the story's opening passages, Douvesne's consciousness is colored by a blurred mélange of reverie and actuality, and Custot describes him as always being "a little bit detached from contact with reality." Even though he entered the military, Douvesne did not do so to serve his patriotic duty; rather, he enlisted so that he could enter a world of escape and adventure, and he was particularly drawn to the navy because he thought the uniforms and military parades would give him a degree of respect that he did not have before. Guy-Emmanuel de Césade, the central opium-using character in "Une âme," one of the stories in Boissière's *Fumeurs d'opium*, was also attracted to the military more by the allure of adventure than the call of duty. Enlisting because of dreams fed by Jules Verne books and a desire to avoid school, Césade, like Douvesne, is a man who sees military service not as a sacred duty to the nation but as an expedient that can propel him on a path to adventure and glory.[43]

In this genre of opium fiction, the seemingly innocent persuasions of a casual acquaintance usually lead the protagonists to first try the drug. Yet because of their dreamy dispositions, the main characters are drawn in by the allures of opium, and soon it evolves from a mere pastime into a habit. Regular use eventually takes its toll on the physical and mental health of these men. Douvesne, when caught smoking opium on board a military ship, commits suicide by jumping overboard instead of enduring a life without the drug, Farrère's opium smoker becomes haunted by visions of the living dead coming back to life in a cemetery, and Boissière's opium adepts become overwhelmed by listlessness, poor health, hallucinations, and visions of ghosts.[44]

In these stories, beyond compromising the mental and physical health of users, opium also makes soldiers unfit for military service because of its effects on their hearts, minds, and souls. Custot, Farrère, and Boissière each took pains to emphasize that their opium-using characters are poor soldiers, not simply because the drug eats away at their strength, but also because it compromises their loyalty to the army and, by extension, to France. In *Midship*, for example, the irreconcilability of opium and service to the nation becomes clear in the story's penultimate scene, when Douvesne's commanding officer reprimands him for setting up an opium den on board. The commander castigates his men for choosing to indulge in a drug that compromises their ability to carry out their duty to the nation. "Your minds, instead of putting them at the service of our Navy, you are letting them go up in smoke," he charges. Opium saps not only the physical strength of users, the officer maintains, but also their moral strength. Thus soldiers must be drug-free to preserve both the health of their bodies and the fervency of their patriotism. They need to stop smoking opium "so that your souls can be purified and your entire existence can be . . . a hymn to France." Opium, in Custot's story, threatens the navy, not only because it makes poor sailors and fighters, but also because it makes poor patriots, men whose loyalties to France are compromised by the allures of the drug.[45]

Other authors went beyond pedagogical speeches to emphasize the spiritual threat that opium posed to the men of the armed forces by compromising their loyalties and identities. In one of the vignettes in *Fumeurs d'opium*, for example, the protagonist begins to identify with his Asian subjects, claiming that as his opium use escalates, he begins to live "à la tonkinoise, eating, smoking, and sleeping on the same mat, more Annamite than the Annamites." In *Black Opium*, Farrère describes a Frenchman's sense of alienation from his homeland and feeling of solidarity with the Asians on entering a Chinese opium den. Extolling the pleasures and deep contemplation that opium opens up for him, Farrère's narrator makes the distinction between opium and Frenchness abundantly clear. "I can better feel brother to Asiatics smoking in Foochow Road," he declares, "than I can to certain inferior Frenchmen now vegetating in Paris where I was born." Opium, he explains, is "a fatherland" to him, one that commands more of his loyalty than any other solidarities he previously felt, including that with the French nation. "The European, the Asiatic, are equal," he explains, noting that opium seemingly "effaced" particularities of race, physiology, and psychology. Instead of the European cities he knows, he comes to prefer the lands of "dreams which are very ancient and very wonderful." Opium, therefore, obliterates the narrator's identity as a French citizen, making him more a timeless and nomadic dreamer than a patriotic servant of the nation in the here and now.[46]

The link between opium and poor citizenship is most striking in Boissière's *Fumeurs d'opium*, as both the sentiments and actions of his opium-using characters make their irresponsibility, their lack of patriotism, and their failure to carry out their duty to the nation abundantly clear. One common theme is the use of opium as a tool of treachery, a drug that Annamite rebels use to dupe French soldiers and make them easy prey for enemy attacks. In one of Boissière's tales, "Le blockhaus incendié," a French soldier serving in Indochina begins taking opium when he befriends an Annamite man who, it is later revealed, is a rebel leading an insurgency against the French. The drug encourages the soldier to "fraternize" with the Asians, creating a dangerously "reciprocal benevolence" that eventually leads him to let his guard down. The soldier's drug-induced honeymoon comes to an abrupt end when one night, while he is in an opium reverie, the Annamite acquaintance who has been supplying his opium leads a rebel charge against the French, and the Asians massacre the soldiers stationed at the outpost. In another story in *Fumeurs*, "Les genies du mont Tân-Vien," Boissière tells the tale of a soldier who falls victim to Annamite disloyalty because of opium; his assistant, who supplies him with the drug, is actually a covert rebel leader. As in "Le blockhaus," the soldier is led astray by going down a path lined with opium and Annamite duplicity, and the rebels eventually murder both him and his family.[47]

In the final of Boissière's vignettes in *Fumeurs d'opium*, "Une âme," opium takes on an even more sinister character in its effects on the Frenchmen who indulge in it. Césade, the story's protagonist, is a young soldier who begins taking opium to calm nervous pains, and quickly becomes addicted while serving in Indochina. The drug does little to cure Césade's ills, but it opens up a new world for him, and soon he is drawn in by its charms and starts smoking more than sixty pipes a day. Though once adventurous, Césade becomes apathetic when accustomed to opium, maneuvering to avoid going on marches and missions so that he can remain close to his opium pipe. Soon he begins frequenting local opium dens and immersing himself in the drug-themed writings of Baudelaire, and he becomes increasingly fascinated by his own imagination and scornful of his fellow soldiers. As his opium habit progresses, so does his distaste for the camaraderie, solidarity, and devotion that are supposed to be the hallmarks of military life. In one scene, Césade's superior makes the case against opium on a moral front, urging him to stop taking the drug so that he can better fulfill his patriotic duty. But instead of being inspired to change, Césade becomes more obstinate in his selfish and snobbish ways. His superior "spewed theories of patriotism . . . battles and assaults, blowing flags and heroisms," Césade muses. Such appeals, he concludes, are "idiotic," the ramblings of a "simple philosophy" that he deems both naive and childish. After all, military service is but a "servile career, ridiculous to those who discovered wisdom, certitude, and the pride of the free and contemplative man in opium." Thus Césade is unfit to serve because he is both unable and unwilling, as he develops an open hostility to the principles of patriotism and solidarity, and rejects the call of duty in favor of the "free and contemplative" gifts that opium offers.[48]

After a confrontation with his superiors, Césade decides to desert so that he can "smoke in peace." Upon escaping, he joins up with a rebel chief who is leading an insurgency against the French, and Césade offers him intelligence on French military strategy in exchange for protection. The decision to betray his country is an easy one for Césade, since as he becomes further entrenched in the opium habit, he grows increasingly scornful of the principles of patriotism. "Faithfulness to the Flag, the symbol of the nation," he sneers as he meets with the rebel leader, "always seemed to my reasonable intelligence as a stupid sentiment." The choice, therefore, is an easy one; without hesitation, Césade joins the rebels and takes up arms against France. Soon thereafter, Césade begins smoking opium with the rebel leader and becomes one of his closest confidants. He also begins to develop a rapport with others in the insurgent camp, feeling particularly close to them because of their shared anti-French sentiments. "The people of France," Césade writes in his journal, have become "our common enemy," a focus of hatred that attracts him to the rebel cause. "I have thought too much to give my love just to men . . . born on one side of the Alps, the Pyrenees, or the sea," he asserts, quipping that "we best leave this to the kids in primary school." For him, any culture or nation can be appealing, as long as it is one that violently opposes the French. "I prefer Bismarck, Goethe, and Confucius to the entire municipality of Paris."

Aligning himself with the cultural icons of France's enemies—Germany and, in the colonial context, Asia—Césade throws himself into the anti-French crusade. The prospect of providing intelligence that could lead to the defeat of his former comrades gives Césade little anxiety; instead, it overwhelms him with a feeling he describes as "joy."[49]

Césade's plans fall through, however, when the French catch him and sentence him to face a firing squad. Even as he awaits his death, Césade continues clinging to opium and its ability to procure intellectual joys and powers as his salvation. Though his former countrymen consider him a victim of opium, Césade maintains that the drug has helped him, not ruined him. "They remain unaware," he writes in his journal shortly before his execution, that he had developed a "subtle intelligence, large and healthy" thanks to opium. Through his final moments, Césade maintains that, with opium, he remains superior to the "simplistic" concepts of nationhood, solidarity, and devotion that are the hallmarks of the good soldier. Opium, therefore, has not only made him a weak soldier and a traitor on the battlefield but, worse, it has turned him into the ultimate anti-patriot and anti-citizen.[50]

An army peopled with such soldiers, clearly enough, could have dire consequences for the French nation. And as a bizarre series of events in the autumn of 1907 showed, the potential for opium to compromise the integrity of French soldiers was more than the stuff of fiction.

Opium on Trial: The Ullmo Affair

The threat of opiate-induced treason stepped out of the annals of fiction and into the reality of the French military on September 9, 1907. It began when Minister of the Navy Gaston Thomson received a letter from an individual claiming to possess secret naval transponder codes and threatening to sell them to a foreign power if he did not receive a payment of 150,000 francs. The author of the letter provided one of the codes in his correspondence and demanded that the authorities accept his offer by acknowledging it through a message in a newspaper. In a second communiqué, the writer lowered his price and instructed the authorities to pay him by leaving an envelope full of cash in the armoire or under the sink of a washroom on a train car in Marseille.[51]

Thomson notified intelligence, which took on the case and established a correspondence with the blackmailer in two newspapers. Agents went to the assigned train, but the drop-off did not pan out. Shortly thereafter, an agent received new instructions, this time telling him to go to a Toulon brasserie, place the money in a waterproof envelope, and leave it in a hole in the wall next to the toilet or in the reservoir of the cistern itself. Once again, the plan failed to yield any results. At this point, agents contacted the thief, telling him that the money was ready to deliver. They arranged to meet just outside Toulon on the night of October 22. Shortly after arriving for the exchange, an agent was approached by the thief, who was wearing a hat and a large pair of chauffeur's glasses to obscure his face. When they were less than a meter apart, the blackmailer pulled out a gun. At this point, the agent jumped on him, threw him to the ground, disarmed him, and fired two shots in the air to call for backup.[52]

The blackmailer turned out to be a twenty-four-year-old Jewish naval lieutenant named Charles Benjamin Ullmo. When interrogated, Ullmo admitted that his plan had involved more than a simple threat of blackmail. He hatched his scheme after reading a newspaper article about a spy who had sold secrets to a foreign power; intrigued by the idea, he proposed a similar deal to a German official in Paris. The Germans expressed interest, so that summer, Ullmo stole secret information from the desk of one of his superiors. When he met with the Germans for the exchange, however, they refused his offer, and at that point Ullmo turned to the idea of blackmailing the French navy so that he could make some money off his plot.[53]

Since Ullmo confessed the facts of the case, there was no doubt as to his guilt. When it came time for his trial, the defense focused on bringing up the question of responsibility: to what degree could Ullmo be held accountable for his actions? Though not denying the gravity of his attempted treason, Ullmo's counsel sought to paint his actions not as the work of a cold, calculating turncoat but rather as the desperate plans of a confused, muddled mind. In the trial, Ullmo's defense attorney, Antony Aubin, argued that Ullmo was a victim of poor heredity, a manipulative woman, and, most importantly, opium. The three combined, he argued, led Ullmo to his attempted treachery.

First, Aubin maintained that Ullmo was a victim of a "psychopathic heredity" that rendered him unable to resist the temptation to blackmail when the idea first entered his mind. Court-appointed psychologists had found that Ullmo came from a compromised family tree that featured eight mentally ill relatives. Turning the logic of degeneration into an extenuating circumstance for his client, Aubin argued that Ullmo had inherited a pathological inability to "resist the solicitation of his appetites." Thus, when presented with the opportunity to turn a profit by spying, Ullmo could not repel the treacherous temptation. A series of emotional traumas—the death of his parents, the difficulties inherent in serving as a Jew in the French armed forces during the Dreyfus Affair, and repeated gambling losses—had left him even more unable to resist the urge to steal the naval secrets. Then in 1905, another factor entered the equation for Ullmo, as he began a liaison with Lison Welsch, a Toulon demimondaine. Ullmo squandered his savings to please Welsch, paying for her to have a servant and a gardener and promising her jewelry and trips to the theater. Soon after starting to see her, Ullmo had squandered his inheritance on her, and he had to start gambling again to get the money he needed to keep her happy. In early 1907, heavy losses at casinos in Monte Carlo ruined him, and he was on the verge of having to give her up, since he could no longer afford to provide her with a luxurious lifestyle.[54]

It was at this point, according to Aubin, that another constitutional weakness—a taste for opium—got the best of Ullmo. Like many of his cohort in the French navy, Ullmo was a somewhat regular opium smoker, first frequenting opium dens in Indochina and Toulon, then smoking regularly with Welsch. Burdened with an already weak mind and worries about keeping Welsch happy, Aubin claimed, Ullmo was, with opium, pushed over the edge of reason and did not know any better when thoughts of treason and blackmail entered his mind. Ullmo's defense even went as far as to shift opium's role from that of accessory to that of facilitator: "His plans for this horrible plot were imposed on him in the midst of opium dreams," Aubin explained, claiming that Ullmo was under the influence of the drug when he devised his plan. Ullmo's recounting of the events seemed to corroborate Aubin's claims, making the correlation between Ullmo's impulse to treachery and his opium use seem direct. According to the defense, it was as if opium created an alternative personality for Ullmo, one driven to betrayal when under the drug's sway. As in the stories of Farrère and Bossière, opium was, according to Aubin, a substance of sabotage that flipped a treasonous switch in the mind and soul of the drug taker. Thus Ullmo was not responsible for his actions; opium was the truly guilty party.[55]

To emphasize the causal link between opium and Ullmo's treason, Aubin tried to depict his client as an incompetent, addicted bungler whose mental state was dominated by an alternation between opium hazes and painful withdrawal symptoms. Ullmo was far from the consummate spy, as he was amateurish in his maneuvers and, as Aubin put it, "devoid of sangfroid." In his contacts with the Germans, for example, he made the mistake of telling them where he lived, even signing his own name and address on one communiqué. Furthermore, Ullmo's plot bore the signs of stupidity, madness, or both: he had considered disguising himself in a velvet wolf suit to go to the drop-off outside Toulon, instead of wearing glasses. And the plans involving bathroom basins and toilets were, in Aubin's opinion, "burlesque" and "crazy." Opium must have played a part, as no serious spy would resort to such amateurish or silly modus operandi. Furthermore, Aubin asserted, Ullmo was indeed an opium addict since he experienced withdrawal symptoms if he did not take opium pills while away at sea. Thus the question, for Aubin, was clear: "Can the experts deny that opium was partially responsible for the plan . . . in Ullmo's brain?" Or, as writers at *Le Figaro* framed it, was the case one of "blackmail or insanity"?[56]

Unfortunately for Ullmo, his state-administered psychiatric evaluation did not support a story that framed him as a victim of opium. Aware of the arguments that Ullmo's defense would bring to the tribunal, the judge in charge of the investigation had asked two medical experts to examine Ullmo and determine how big a role "the prolonged use of opium and moral depression it caused" played in his crimes. In their investigation, the doctors found that Ullmo may have suffered from a degree of mental distress, but it was not nearly as dramatic as the defense team claimed. Heredity, they concluded, was not a legitimate excuse for Ullmo's actions, especially since an examination showed that Ullmo was in good physical and mental health and was blessed with a mind that was "remarkable in its precision, rapidity, and richness." What is more, naval evaluation reports made no reference to Ullmo's having a tainted physical or psychological makeup, instead concluding that he was both "intelligent" and "observant." What did come across in his naval record, however, was that Ullmo had a bad attitude. Despite his apparent talents, superiors noted, Ullmo's abilities were mitigated by an arrogant streak that made him more a slacker than a team player. He was a man who failed to "take his service seriously enough," as he was too "uninterested in his duties and his work as a sailor" to excel. "He appears [to be] . . . a man of little will," they concluded, "lacking ambition and courage."[57]

Such traits, the doctors argued, accorded well with the vices that would bring Ullmo down. Living beyond his means during his time with Welsch was a result of his self-indulgent desire to live a vain and pleasurable existence. A penchant for gambling resulted from a constitutive fault in his character. Furthermore, his personality gave him the prototypical character of an individual drawn to the psychotropic allures of opium, even when not under the influence of the drug. As one of his naval evaluators wrote, Ullmo seemed to believe "that he has nothing but rights, intentionally forgetting his duties." Thus opium use was not a cause of Ullmo's failed sense of duty but rather a consequence of it, part of a larger pattern of selfishness and moral weakness. "The opium habit needs to be, for Ullmo, as for all opium addicts," his examiners concluded, "interpreted as a *sign* of a lacking will and an abnormal appetite," not the author of them.[58]

Even if Ullmo had been poisoned by opium, as the defense claimed, the drug could not have played the role in his treasonous actions that Aubin claimed it did. The defense had argued that Ullmo was always under the influence of opium when devising and carrying out his plan and that, when off the drug, "he thought of [the plan] no more." But when feeling opium's psychoactive effects, the medical experts explained, users tended to feel "detachment" from their problems. It would defy scientific logic, therefore, for Ullmo to have been thinking about his financial troubles while in a drug-induced haze. Furthermore, Ullmo claimed that he was intoxicated by opium when he carried out much of his plot, yet the drug, according to the scientific thought of the day, engendered a "psychological state that itself is unfavorable towards initiative and contrary to action." Consequently, opium users were unlikely to break the law when under the influence; they were more likely to sit in a stupefied trance than to take any criminal action. Thus Ullmo's assertions were paradoxical and defied scientific understandings of opiates' effects on behavior. "Ullmo deployed voluntary intellectual activity that we declare to be incompatible with the state of . . . apathy and inertia of the opium addict," the doctors concluded. The drug, therefore, was "not at all a determining factor" in the "psychogenesis" of Ullmo's treason. After two hours of deliberation, the court agreed and sentenced Ullmo to military degradation and a life of exile in Guyana.[59]

Opium, ultimately, was exonerated. It did not facilitate Ullmo's treachery and was not a mitigating cause. The trial did, however, make the connection between poor citizenship and opium users manifest. Though it was not found to have directly caused Ullmo's betrayal, as in the fictional narratives of Farrère and Boissière, the drug reflected and brought to the forefront a selfish and treasonous streak in Ullmo's character. Opium did not have the power to create treasonous thoughts, but it held a specific appeal to those whose loyalty to the nation was suspect. The drug did not instigate Ullmo's treacherous impulses, but rather, as one commentator claimed, it "embroidered" a design of treachery on a preexisting pattern of poor character and citizenship. Over the course of the trial, Ullmo became the public face of opium and by far the most famous (or infamous) public figure that made the connection between the drug and disloyalty clear. The case was covered extensively in the French press throughout late 1907 and early 1908, and the media coverage helped further link Ullmo, treason, and opium in the public eye.[60]

Ullmo's reputation as the antithesis of the virtuous and devoted Frenchman was uncontested. According to anti-Semitic commentators such as Léon Daudet, Ullmo's disloyalty stemmed from the fact that he was Jewish, which gave him an "aptitude for treason" that was "common among the Israelites." For Daudet and his fellow anti-Dreyfusards, Ullmo's case was a natural follow-up to the Dreyfus Affair, a simple case of Jewish treachery manifesting itself in the navy instead of the army. Yet, even outside the anti-Semitic ramblings of the anti-republican right, Ullmo's case spoke volumes about the patriotic health of the nation and the place of opium users within it.[61] This became abundantly clear in June 1908, when Ullmo was stripped of his military rank and sent into exile in a public ceremony held in Toulon.

Ullmo's degradation had the trappings of a nationalistic purification ritual more than a mere act of military discipline. A massive crowd, estimated to be in the tens of thousands by some observers, showed up at daybreak to witness the event. Arriving armed with picnic baskets and a raucous arsenal of insults for the traitor, the crowd, according to *Le Matin* correspondent Leo Gerville-Reache, was in a patriotic and celebratory mood, looking forward to the degradation as they would a military review or a presidential speech. After hours of anticipation, Ullmo was ushered into the public square and, according to Gerville-Reache's account, he remained stoically indifferent to the mob. Faced with thousands of jeers and insults, Ullmo did his best to remain distant, perhaps, speculated Gerville-Reache, because of opium. Struck by Ullmo's lack of emotion, the reporter wondered "what fantastic dreams conceived of in the heavy vapors of opium" filled Ullmo's mind as he faced the crowd. As it had apparently done when he devised his plot to betray France, opium now seemed to help Ullmo isolate himself during his degradation. Yet, despite his disengagement, Ullmo's public and ceremonial humiliation served a purpose that was both practical and ideological. Joyously and enthusiastically stripping Ullmo of his military rank, the people (represented by both the military and the crowd) exacted their revenge on the traitor, symbolically isolating him—and, by extension, the opium habit he had come to represent—from the rest of the nation.[62]

Providing a real-world example of what opium's relationship with loyalty and citizenship could be, the Ullmo Affair confirmed, rather than assuaged, French anxieties about the place of opiate users within their society. Given that opiate habits seemed to be on the rise and penetrating all classes and demographics, the example of Ullmo was a particularly frightening one—especially for a nation already rife with fears of degeneration and national decline following the Franco-Prussian War and the Paris Commune, and anxious about another pending conflict with Germany. Both a sign and an instigator of national decline, opiates and opiate users seemed to threaten the physical health of the nation and its political and ideological well-being.[63] Concerns about the relationship between opiate users and the nation, elaborated in fiction and in the headlines, would continue to color French attitudes and approaches to opiate use when more stringent controls over the substances and those who used them were put in place in the opening decades of the twentieth century.

A Tale of Two Drug Problems

As the twentieth century began, the perceived threats posed by opiates were drastically different in Britain and France. Even though per capita levels of opiate use were probably higher in Britain than in France, British writers maintained that problematic use and addiction were marginal phenomena, while the French believed that opium and morphine habits were becoming increasingly widespread across the socioeconomic spectrum. Yet the difference was not simply one of depth and breadth, but also one of meaning. The British and French had divergent understandings of what opiate habits could do to the physical health of citizens—and to the essence of citizenship itself—if allowed to fester and spread, leading each country to develop its own distinct form of anti-narcotic nationalism.

In Britain, fears about the potential consequences of unfettered opiate use centered around ideals of productive citizenship that converged with preexisting Sino-phobic undercurrents. When examined more closely in the 1890s and the first decade of the 1900s, these fears proved generally unwarranted, and the opiate threat did not seem immediately menacing to most Britons. In France, however, fears about opiate use and its relationship with republican citizenship, as the French understood it, seemed real and were magnified by the Ullmo Affair. Not only was opiate use apparently becoming more prevalent, but it also had a more menacing character, affiliated with crime, poor citizenship, and treason.

Thus the limited concern about opiate habits in Britain and the relative panic emerging in France resulted from differences in perceived drug demographics and in drug discourses. These patterns of thinking about opiates, and the dangers they posed, would continue to play significant roles as Britain and France forged their more comprehensive drug control regimes in the first two decades of the twentieth century.

CHAPTER THREE

The Era of National Narcotics Control

The Drug Wars Begin

Drug addiction is, at bottom, a matter of drug supply.

—*Malcolm Delevingne*

Although a global system of narcotics control would not come into effect until the 1920s, many nations—including Britain and France—instituted new domestic regulations to limit the consumption and use of opiates in the first two decades of the twentieth century. International forces and universal concerns about public health were at play, but nations also had their own reasons for instituting tighter controls over narcotics, ones that influenced both the timing and the character of their drug control initiatives. In Britain, the direct impetus for opiate control was fear of economic problems, while in France, tighter regulations were put in place to address concerns about what opiate use could do to the political and ideological health of the nation during a time of war. As in the development of each country's anti-narcotic nationalism, the push for more enthusiastic state action to control opiates reflected broader conceptions of what defined the nation and opiates' potential to destroy it.

The international opium agreements of the pre–World War I era helped set the terms of the drug control enterprise, but distinctly national understandings of what the "drug problem" was remained intact well into the interwar period, shaping how Britain and France set goals, formulated policies, and undertook the task of limiting the flow and consumption of opiates in their territories. Thus, even though the war on drugs was a global one waged mostly in the twentieth century, it remained largely influenced by national styles of thinking about opiates that had developed at the end of the nineteenth.

The International Context

Around the time that tighter narcotics control measures would take effect in Britain and France, a movement to restrict the use of opiates to medical purposes was developing throughout the world. By 1908, many industrialized countries had issued laws and regulations to control the commerce in and use of opiates. This coincided with the development of an international campaign to assist China as it tried to check the spread of the opium habit. In 1903, Britain and the United States agreed to cease morphine shipments into China, and in 1906, the Chinese government issued edicts calling for the suppression of domestic opium cultivation and use. Driven partly by a humanitarian impulse to help China and partly by a desire to curry favor with the Chinese for commercial advantage, several nations—including Britain and France—agreed to help. The British pledged to decrease imports of Indian opium into China by ten percent every year, beginning in 1908, and the French promised to shut down opium dens in their concessions on Chinese territory if other colonial powers did the same.[1]

European initiatives to support China were not as enthusiastic as those of a new colonial power in the Far East—the United States. Shortly after gaining control over the Philippines in the Spanish-American War, U.S. officials and clergy tried to curb the opium habit in their newly acquired territory, only to be frustrated by the prevalence of smuggling throughout the island chain. Furthermore, as the only colonial power in the region without a foothold on the Asian mainland, the United States was particularly eager to assist Chinese anti-opium efforts in order to secure favorable trade agreements. Based on its experience in the Philippines and its knowledge of China's history with the drug, the United States believed that the only effective way to combat the opium evil in the Far East was through global cooperation. Eager to take action, the United States called for an international conference to address the drug problem in East Asia, and it organized a meeting that convened in Shanghai in February 1909.[2]

Despite the United States' best efforts to forge a consensus on how to place more stringent controls on opium, the representatives at the conference were unwilling to jeopardize their own countries' economic interests. Nonetheless, the nations present at the Shanghai Commission, with the exception of Portugal, agreed on nine resolutions confirming the necessity of an international effort to help China suppress the opium habit. More important for the future of participating countries' domestic drug laws was the commission's call for each nation at the conference to institute tighter internal controls on the circulation and use of opiates. The signatory powers agreed that governments should take measures to suppress opium smoking, that each country's opium regulations should aim toward "progressively increasing stringency," and that each nation should take steps to regulate not only opium but its potentially addictive derivatives. The commission also agreed with the United States' call for the control of opiate exports, not just to China, but also to any country that had laws regulating the drugs, thus establishing the principle that drug control was to be an international undertaking.[3]

The Shanghai Commission's resolutions, however, were only of an advisory nature, and they set out no specific legislative or administrative roadmap for signatory powers to follow. Consequently, few of the nations that signed the Shanghai accord acted to bring their domestic policies in line with the conference's recommendations. By the end of 1909, the United States began calling for another conference to draft a treaty that would lay out more definitive steps toward an international drug control regime. The meeting convened at The Hague in December 1911. Many details concerning the aims and mechanics of international drug control proved to be major stumbling blocks as countries worked to draft an agreement. Some nations (France, Germany, the Netherlands) were reluctant to sign an agreement that would compromise their sovereignty by requiring them to alter their national legislation, regulate drug manufacturers, or allow foreign agents to search their ships. Others (Britain, China) still did not agree with the U.S. delegation on how to distinguish "medical" from "recreational" use, thus complicating the task of discerning legal opium from contraband.[4]

The biggest stumbling block at The Hague, however, was the effect that drug control would have on commercial and economic interests. Due to the profitability of colonial opium manufacture and monopolies throughout Asia, many powers were reluctant to sacrifice revenues gleaned from the opium trade. Furthermore, several countries, particularly Britain, were wary that if they agreed to stop producing and shipping opiates, others would step up operations to profit from the resulting vacuum in the narcotics marketplace. That two major players in the international drug business—Turkey, a major producer of raw opium, and Switzerland, a principal manufacturer of morphine and heroin—were not at the conference made this fear seem all the more legitimate.[5]

Thus, to institute effective control, any international agreement would have to meet two key prerequisites. First, it would need the signatures not only of the nations that drafted the accord but of all countries involved in the opium trade. Otherwise, the opiate traffic would not be checked but would simply move to wherever it could operate without restriction. Second, such an agreement could not simply target raw opium; it also needed to control the drug in its manufactured and synthetic forms. Consequently, it was not just the opium-growing parts of Southern Europe, the Middle East, and Asia that would need to voluntarily limit their business; the major pharmaceutical manufacturers in western Europe and North America would also have to do so. Opiate control, therefore, would be both difficult to coordinate and costly for farmers and industrialists throughout the world.

Given the drawbacks of comprehensive regulations, participants at The Hague were not eager to commit themselves wholeheartedly to the narcotics control effort. The agreement that emerged out of the conference in January 1912 did little to create an effective drug control regime, as representatives agreed to sign an accord that was both vague and noncommittal. Instead of establishing policy goals or guidelines, the convention participants simply called for powers to "enact effective laws or regulations" or "use their best endeavors" to institute controls "as soon as possible," without defining what such measures would be or a target date by which they should be taken. Moreover, the convention did not set out a uniform penal regime or define what should constitute a drug-related offense, though it did make some recommendations. The convention's biggest weakness lay in its provisions concerning ratification. To assuage concerns that drug control would be ineffective unless all countries involved in the drug trade (not just those present at the conference) were in agreement, the treaty stipulated that the protocols—indefinite though they were—would not take effect until the majority of the drug-producing world signed the accord. In Article 22, the convention listed thirty-four countries that were not present at the conference but would need to sign the treaty for it to become operational. By the end of 1913, nine of these countries still had not signed the agreement, many of which were opium-growing states in southeastern Europe that were too preoccupied with the Balkan Wars to consider the treaty. In June 1914, the powers met at The Hague once again, calling for all of the countries to sign the accord by the end of the year. Before the ratification process could be completed, however, it was derailed by the beginning of World War I in August, and the task of international drug control was put on the backburner until the end of the conflict.[6]

Although it did not take effect until after World War I, the Hague Convention was nonetheless a big step toward the creation of an international drug control regime, as it established a framework that shaped both the form and content of subsequent national drug control initiatives. It laid out definitions of what constituted the major opiates—"raw opium" (coagulated juice obtained from *Papaver somniferum*), "prepared opium" (i.e., "smoking opium," raw opium dissolved, boiled, roasted, and fermented), morphine (the alkaloid $C_{17}H_{19}NO_3$), and heroin (diacetylmorphine, $C_{21}H_{23}NO_5$)—and cocaine ($C_{17}H_{21}NO_4$). The convention also specified which medical preparations that included these drugs should be regulated, recommending that any medicines with more than 0.2 percent morphine, 0.1 percent heroin, or 0.1 percent cocaine should be more tightly restricted. In so doing, it set an internationally accepted definition of which substances were to be controlled, and at what levels. These standards would guide many countries as they began drafting their own drug control legislation over the next five years. Furthermore, in spite of its wishy-washy language and difficult ratification procedure, the Hague Convention set out what the major goals of drug control were to be: the limitation of exports and imports, the restriction of access to the drugs for anyone other than doctors and pharmacists, the repression of opium smoking, the proper labeling of narcotics, and the suppression of smuggling.[7]

In the 1920s, the goal of implementing a more rigorous and binding international control regime was finally realized. As many countries learned during the war, smuggling was a problem not just in the Far East but also in Europe and North America. The United States, Britain, and China advocated making signing an international drug agreement a condition for peace; as a result, Article 273 of the Versailles Treaty compelled defeated countries that had been reluctant to sign international drug agreements—most notably Germany and Turkey—to ratify and implement the Hague Convention. Article 23c of the Covenant of the League of Nations authorized the new organization to oversee the execution of the Hague agreement, and until its dissolution, the League was the central organ for international drug control. To cut down on smuggling, the League suggested that countries adopt an import certificate plan, under which nations would not allow narcotics to be sent abroad unless exporters were provided with a certificate from authorities in the destination country affirming that they knew the drugs were being sent and guaranteeing that they would be used only for medical or scientific purposes. The 1925 Geneva Opium Convention instituted the import certificate system and created a new body, the Permanent Central Board, to track the international trade in narcotics, collect statistics on the growth, manufacture, and trade of drugs, and gather intelligence on smuggling. During the 1920s, the League also worked to estimate the total global requirements of opiates for medical purposes, with the hope of restricting production and cutting off supplies that fed the black market. In 1931, an international agreement limiting the manufacture of narcotic drugs to the world's medical and scientific needs was ratified, and a new organization, the League's Drug Supervisory Body, was established to determine what the global need for narcotics was and to set limits on how much each country could produce annually. Thus, by the eve of World War II, the comprehensive international control scheme envisioned by the United States some thirty years earlier was finally taking shape.[8]

There were, however, still significant variations in drug policy across borders. The 1925 and 1931 agreements focused more on the large-scale commerce in narcotics and limiting drug production than on domestic strategies for controlling those who sold or used controlled substances. Agreements concerning these issues would not become operational until well into the postwar period. Thus international accords had only a limited impact on how individual nations addressed their drug problems at home during the period from 1900 through 1925. Given that no convention had come into force before World War I and that the guidelines set out at The Hague were relatively hazy, national authorities controlled opiates as they saw fit during this time. Consequently, early domestic control initiatives, though influenced by global developments, were shaped as much by national motivations specific to each individual state as by universal concerns about public health and morality. In Britain and France, this was definitely the case, as tighter controls over opiates and those who used them emerged as a set of domestic policies designed to address a particular set of national concerns.[9]

Britain in the Age of National Opiate Control

The Prewar Context

At the time of the Shanghai and Hague conferences, there were few restrictions on opiates in Britain. Though opium was considered a potentially "poisonous" substance, until the twentieth century it was simply regulated as an ingredient in medical preparations. Under the 1868 Pharmacy Act, the only restrictions on the sale of the drug were that only qualified vendors could sell it and that it needed to be properly labeled. The 1908 Poisons and Pharmacy Act instituted tighter controls by reclassifying opium as a Schedule I poison, meaning that sellers had to either know or be formally introduced to anyone who purchased it and that merchants needed to keep records of their opium sales on a poisons register. Nonetheless, opium was still available in unlimited quantities, and no medical prescription was required to buy it from a retailer. As barrister Wippell Gadd quipped in 1911, "the sacred right of British sub- jects to obtain what poison they wish" was ensconced in law. When it came to the export of opium, the British control regime was also loose. Manufac- turers that produced morphine, heroin, and other opiates were not subject to as much excise control as were producers of alcohol, and customs officials had no procedures in place to limit opiate exports, other than a requirement that companies document the amount of drugs they shipped overseas. Fur- thermore, the penalties for violating controls—fines of five pounds for the first offense and ten pounds for subsequent violations—were minimal, and not severe enough to discourage individuals, particularly habitual users or addicts, from ignoring the regulations.[10]

Evidence from the British National Archives suggests that officials in Whitehall were not overly concerned with the use of opium on domestic soil at the time of the Shanghai conference. "There are no regulations in force in this country applying to opium dens," they admitted as they prepared for the meeting. Although some members of Parliament had proposed legislation to create institutional treatment options for drug addicts, their bills did not lead to any changes in policy. There was little need to repress recreational drug use, British officials reasoned, since, as far as they knew, it "could hardly be said to exist," and opium smoking was "rather on the decrease than oth- erwise." Where opium was a problem, it seemed that narrowly tailored local initiatives, such as the London County Council's 1909 by-laws prohibiting opium smoking in licensed seamen's boarding houses, were enough to ad- dress the issue.[11]

Nonetheless, after signing the accord at The Hague, the British wanted to fulfill their commitments, vaguely defined though they were. From March through July 1914, an interdepartmental committee met to consider what measures should be taken to comply with the Hague Convention. Given that a new drug regime would require changes on both the domestic and inter- national levels, the committee brought together officials from an array of ministries—from those concerned with colonial matters (India Office, Colonial Office) to experts in international affairs (Foreign Office), the architects of domestic policy (Privy Council Office, Home Office), and representatives of economic interests that would be touched by the new legislation (Board of Trade). The committee considered many actions, such as mandating Home Office licenses for commerce in opiates, requiring prescriptions for medicines with opiate levels exceeding one percent, instituting tighter regulations over wholesale opiate transactions, creating a licensing system for opiate exports, banning shipment of the drugs by mail, and prohibiting opium prepared for smoking. A major thrust of these proposals was to create a system in which opiates could be tracked, from the time they were manufactured or entered British territory to the time they were consumed for medical purposes or sent abroad.[12]

Though it deliberated tighter restrictions, the committee ran into two major problems, both related to something the British held dear: commerce. First, customs officials feared that attempts to regulate the comings and goings of narcotics would be ineffective, tedious, or both. Merchants, customs officials argued, could easily prevent the detection of drug shipments by not entering narcotics on their import/export forms, sending them concealed in other goods, or simply declaring them as "medicines" without specifying their contents. Moreover, customs was powerless to stop the drugs from being picked up by individuals other than designated consignees once they reached foreign ports. The process of ensuring that drugs did not fall into the wrong hands overseas, the Customs Office claimed, would be a "very clumsy" one. Given the amount of correspondence it would require with foreign agents, surveillance of controlled substances would eventually drown customs agents in a sea of paperwork. Consequently, customs officials concluded, no matter what they did, it would remain "a comparatively easy matter to smuggle opium either inwards or outwards . . . without detection."[13]

The second hurdle for drug control was the same concern that caused problems during the negotiations at Shanghai and The Hague—that of national economic interest. Given that many other manufacturing countries had not yet signed the Hague Convention, the Board of Trade maintained that British efforts at international control would be thoroughly "ineffectual" and do little to address the worldwide drug problem. All that being the first country to implement the convention would accomplish, they warned, would be the placement of British manufacturers in an "unfair position of inequality" relative to overseas pharmaceutical companies. Thus, "while not materially assisting countries desiring to control import," limitations on British drug production would "transfer trade and manufacture from the United Kingdom to countries free from regulation."[14] As such, Britain had little to gain from restricting trade in opiates, but plenty to lose.

Given these concerns, the interdepartmental discussions of 1914 remained more exploratory than decisive and, as the Home Secretary would later term them, of a "purely tentative character" throughout. A few weeks after the committee's July meeting, World War I erupted, and the ministries engaged in the drafting of Britain's potential narcotics control regime became occupied by the more pressing matters of the war effort and national defense. A handful of medical researchers continued to push for legislation to check the spread of addiction, but the calls of concerned doctors generally fell on deaf ears in Whitehall until the middle of 1916. Not until the free availability of opiates began to threaten trade, the concern that had held the British back from passing narcotics control legislation, were the authorities driven to take action and restrict access to the drugs.[15]

Drugs and the Defense of the Realm

Evidence from the British National Archives indicates that the loose controls over opiates did not become problematic for central government officials because of abuse by British citizens. Rather, the government was compelled to take action because the drugs' open availability made Britain a center of the international illicit traffic.

Britain became a favorite point for opium traffickers because of simple market economics: it offered an ample supply of the drugs, and users in other countries provided an increasing demand for them. Many other industrialized nations had already taken action to restrict the availability of opiates before World War I, but Britain had not. Consequently, it was easier and less costly to procure opiates from wholesalers and pharmacists in Britain than elsewhere, while in countries that had already instituted controls over opiates, the drugs quickly became scarce, expensive, and thus lucrative for underground dealers. As a result, black markets for opium emerged not only in China but also in North America, continental Europe, and Australia around the time of the Hague Convention. From the perspective of the smugglers who oversaw the contraband drug trade in these countries, the ideal source for opium would be a place that met two key conditions: accessible and inexpensive stores of opium, and convenient access to international ports through established shipping routes. With its relatively loose opium control regime and commercial interests all over the globe, prewar Britain was an illicit drug dealer's dream—as borne out by the speed with which large-scale Britain-centered drug rings emerged between 1910 and 1915.[16]

In February 1916, one of Britain's main shipping lines, Alfred Holt & Co., filed a formal complaint to the government, outlining the troubles that opium smugglers had caused them. Every time a crew member aboard one of their vessels was caught smuggling opium by foreign customs authorities, not only the individual trafficker was subject to legal sanctions, but so was the shipping company he used to transport the contraband. "The firm has in consequence been frequently compelled to pay fines or dues," Holt complained, and had suffered tremendous "discredit" to its reputation because it had become "concerned, against [its] will, in a traffic which they [Holt] know to be illegal and believe also to be immoral." Because of past implication in trafficking operations, Holt's ships had become subject to significant delays: they were now required to search their own crew members and to wait before unloading goods in foreign ports. Even worse, the sanctions the company faced had the potential to devastate business, as Holt had to pay heavy fines and risked being banned from major ports. If such expulsions were to happen, particularly in China, Holt warned, this would damage not just the company's interests but those of all who had a stake in British trade.[17]

Even if shipping companies could avoid severe punishment from foreign authorities, Holt officials claimed that their company had become involuntarily, and unjustly, subject to "great anxiety, trouble, and expense" because of illicit trafficking. To avoid draconian penalties, shipping lines were compelled to "assume the functions of an enquiry and detective agency, as well as those of shipowners." But in spite of great time and expense devoted to rooting out drug smugglers, the problem remained endemic; between July 1914 and December 1915, notwithstanding a royal proclamation banning opium exports during the war, nearly twelve tons of illicit opium were seized on Holt's liners, and this was probably just a fraction of the contraband sent on its ships. Holt needed help in its campaign against opium smugglers, and the entire economy would suffer, the company argued, if the government did not step in. "The suppression of the trade . . . [should] not be abandoned to private effort," Holt reasoned, since both "public interest" and "public honor" were compromised by the use of British vessels for opium trafficking. In a time when commerce was already restricted due to the war, the British economy could scarcely afford the potential costs of letting drug rings continue operating in Britain with such ease. Furthermore, since it was understood that a major responsibility of the British state was to facilitate trade, Holt concluded that it was incumbent on the government to take action and mobilize a defense against the drug-smuggling menace. "It is only just that the State, and the State alone," Holt asserted, "should be called upon to repress" the illicit drug trade.[18]

Having gained intimate knowledge of the difficulties involved in stopping smugglers, Holt suggested some potential changes to the law that could facilitate more effective surveillance. Existing regulations, the company claimed, were grossly inadequate: unless a smuggler was caught with opium in his hands, it was impossible to take legal action against him in Britain. Even when traffickers were prosecuted, the penalties levied on sailors were not severe enough to discourage smuggling, and syndicates were able to provide funds to help their henchmen out of legal trouble. Though wartime provisions allowing for the detention and deportation of foreigners helped in the prosecution of some Chinese dealers, the legislative machinery in place did not address the broader problem that, according to Holt, made Britain-based smuggling possible: that drugs were so easy to get on British soil. "So long as it is possible for the principals in England to acquire large quantities of opium," Holt explained, "the joint operation of laxity in the United Kingdom and severity abroad make illicit traffic easy and profitable." To fix the problem, the company concluded, "the evil should be attacked at the root" so that the contraband trade could be stopped "in its early stages." If traffickers could be denied access to the British pharmacists and wholesalers who supplied them with opium, the entire contraband enterprise would be choked off before any drugs reached British ships.[19]

To facilitate the institution of a more effective control regime, Holt suggested five changes that it believed would help suppress the illicit opium trade in Britain: (1) requiring that all opium imported into the country be registered with the government, so that the quantity and location of all opium on British soil would be known; (2) limiting wholesale purchases of opium to qualified pharmacists, medical professionals, and university researchers who used the drug for medicinal purposes; (3) prohibiting opium exports, except with a special license; (4) authorizing police to search individuals suspected of violating opium regulations and seize all opium in their possession; and (5) imposing heavy penalties on wholesale dealers who violated opium regulations.[20]

Within months of hearing the complaints and suggestions put forth by

Holt, the government moved to take action. When an interdepartmental conference convened in June 1916 to consider Holt's concerns, representatives from the Colonial Office, Home Office, Foreign Office, Board of Trade, and Board of Customs all agreed that opium smuggling indeed posed a grave threat to British commerce. With trade already constrained by the war, the committee agreed that "any unnecessary delay" to shipping caused by the search or expulsion of British liners would be "prejudicial to national interests." Moreover, given that opium was a key ingredient in medical preparations, the committee concluded that the maintenance of adequate stores of the drug for military needs was critical, since smuggling could endanger supplies needed for the battlefield. Holt's case had clearly been a convincing one, and the main question facing the committee was not so much whether new drug control measures were needed but what form they should take. As the Colonial Office summed up in its report on the meeting, the main goal of the new measures would be to prevent "the smuggling of prepared opium from this country to America and the East." Preventing Britons' recreational use of opium, it seems, was a secondary concern in the initial drive toward drug control in 1916. More pressing matters—the need to maintain adequate stores of medical supplies, and threats to the economic health of the nation—sparked the renewed push toward opium control.[21]

The committee agreed with Holt that the most effective way to cut down on international opium trafficking would be to limit domestic availability of the drug. Even before the Holt memorandum, the British authorities had taken steps in this direction, establishing a framework for controlling the flow of other psychoactive substances in an attempt to limit their use by servicemen. A series of administrative measures to limit access to alcohol during the conflict were put in place shortly after the outbreak of hostilities. Regulations issued under the Defence of the Realm Act (DORA) restricted the sale and distribution of alcohol, and a new Central Control Board governing the liquor traffic was established. For other drugs, however, existing legislation and regulations were not strong enough to prevent them from reaching the front lines: the legislation was written too narrowly, it did not give police the rights of search and seizure in suspected drug cases, and its penalties were not severe enough to discourage prospective dealers. DORA provision 40A, which prohibited the supply of narcotics, sedatives, and stimulants to members of the armed forces, was somewhat more successful, as it helped secure the arrests of a few individuals who were dealing cocaine to soldiers. Nonetheless, many in the military believed that cocaine use in the armed forces would remain a problem because the regulations did not go far enough. "So long as the civilian can get it," Sergeant Francis Lloyd wrote to the London police in July 1916, "there are no real means of preventing the soldier from getting or being given it." The only way to keep members of the armed forces from receiving cocaine, military authorities believed, would be to prevent the general population from procuring it. To keep opium away from smugglers, the interdepartmental committee reasoned, a similar strategy would need to be adopted; by restricting access to opiates for everyone, the government could ensure that the drugs would not make their way into the hands of smugglers.[22]

To overcome the weaknesses of existing regulations, the interdepartmental committee decided to integrate the suggestions that Holt and military officials were making into the control mechanisms set up under DORA. By utilizing DORA to tighten restrictions, the committee reasoned, the government could bring opium control under the umbrella of regulations governing national defense, thus giving police the powers of search and seizure, as well as the latitude to use force if necessary. Furthermore, if integrated into the war effort as part of DORA, strict controls over opium would become part of the broader and uncontested task of winning the war and defending the homeland, and thus less likely to provoke significant opposition.[23]

On the specifics of the new restrictions, the committee agreed with Holt and the military authorities that significant changes would be necessary if control over opiates was to be effective. "The only method which [offers] a prospect of success in dealing with illicit transactions," the committee concluded, was to "control the raw material from its first entry into the United Kingdom" until it was consumed. Limitations on who could import opiates, whom they could be sold to, and who in the general public could purchase them from retailers and pharmacists would need to be put in place. In addition, since it was difficult for police to establish whether drug transactions took place legally or not, the new regulations needed to expand the definition of what constituted a violation. "It will be of no value merely to deal with illicit sales," Scotland Yard warned. "It is essential, if the problem is to be seriously grappled with, that . . . unauthorized possession . . . [should be] punishable." Since possession was all that authorities could usually establish, law enforcement officials reasoned, a provision defining illegal possession of opiates as a prosecutable offense could make the task of nabbing traffickers much more practicable. Police also wanted the power to arrest individuals for even attempting to deal drugs, so that peddlers could be prosecuted without necessarily being caught in the act of making an illegal sale. By creating an offense for illegally hawking opiates and also making illegal possession punishable, the new act could simultaneously make the work of underground dealers more difficult and the task of law enforcement easier. To help more effectively prevent cocaine from reaching soldiers, the committee crafted the provisions so that they would control both opiates and cocaine.[24]

The new regulation was issued as an appendix, announced as DORA provision 40B, at the end of July 1916. DORA 40B limited access to opium and cocaine from the time they entered British territory, stipulating that only individuals with a license from the Home Office could import them. Importers, in turn, could supply the drugs only to "authorized persons," a group that included licensed medical practitioners, dentists, veterinarians, licensed pharmacists, and wholesalers, importers, and exporters who obtained a permit from the Home Office. All transactions involving the drugs needed to be entered in a special opium/cocaine register that designated the date of sale, name and address of the buyer, and amount of drugs sold. The register was subject to inspection by Home Office officials during business hours, and the records had to be meticulously kept; any attempt to alter entries constituted a violation of the regulation. If the authorities were not satisfied with the precision or neatness of the register, DORA 40B empowered them to withdraw permission to buy or sell the drugs. Thanks to these provisions, the government, as the Home Office explained, could now know how much of each drug "[is] coming into this country, and where it is going to" at all times. In addition, the system would facilitate the detection of pharmacists or wholesalers who were supplying street dealers or smugglers, as any unusually large drug transactions would come to the authorities' attention when they checked the registers.[25]

Retail sale of the drugs was severely curtailed by DORA 40B. The regulation stipulated that cocaine could be distributed to individuals other than "authorized persons" only through a medical prescription, that prescriptions could not be repeated, and that the drug could be bought or sold only in clearly labeled receptacles. Anyone found in possession of cocaine who could not prove that he was either authorized to have it or had procured it with a medical prescription would be guilty of violating the regulation. For opium, the controls were even stricter: it became illegal for anyone other than "authorized persons" to be in possession of the drug under any circumstances. To facilitate enforcement, the regulation also stipulated that it was an offense not just to sell, give, procure, or supply opium and cocaine but also to "offer" to do so. Consequently, it would be much easier to press charges against illicit dealers, even if they were not caught in the act of making a sale.[26]

Even though the availability of raw opium and cocaine was severely limited, the toughest controls were reserved for opium prepared for smoking—the substance at the heart of the contraband trade for smugglers. Such opium was effectively banned by DORA 40B, as the regulation made it an offense to prepare raw opium for smoking, to be present when opium was being prepared for smoking, to sell smoking opium, to possess it, to manage or attend an opium den, or to possess any pipes or tools that could be used for the preparation or inhalation of the drug. Thus, even more than for cocaine and raw opium, the illicit trade in smoking opium was to be curtailed by keeping the drug away from the black market. Now that the drug would no longer be allowed in the British metropole, trafficking would theoretically become extinct thanks to DORA 40B. Smuggling the opium prepared for smoking out of Britain would be impossible, the logic went, if it was not allowed in the country in the first place.[27]

Unlike the penalties for ignoring the Pharmacy Acts, the punishments for defying DORA were severe. Violators could be given up to six months in prison (with the possibility of hard labor) and a fine of up to a hundred pounds. With increased police powers, up to ten times the earlier fine, and the possibility of prison time, the authorities hoped the new law would either put Britain's drug traffickers behind bars or scare them away from the underground drug business altogether.[28]

Subsequent regulations, issued after DORA 40B, aimed to further choke off supply lines for the illicit drug market. In particular, the Home Office made a point to issue opium and cocaine permits to just a handful of major corporations; by keeping drugs in the hands of just a few businessmen, the government not only made the task of tracking the substances easier but ensured that the companies who dealt in the drugs had a "position or reputation to lose" if found in violation of the law. Well-established businesses that had a significant stake in maintaining their good names and the good graces of the government, officials reasoned, were less likely to be tempted by the allures of the illicit market.[29]

The control of opium and cocaine could have involved many regulatory agencies, but law enforcement took the lead in the administration of DORA 40B. As the authorities had envisioned when they drafted the provisions, police were given the most responsibility in carrying out drug investigations, particularly in London's West End and the Chinese districts of port cities.

Customs officials played a minor role in the enforcement of opium and cocaine regulations. If they detected a drug law violation, they were to immediately report suspicious findings to the local police, who would then come to the docks and take over the case. After a 1919 court ruling blocked the Board of Trade from detaining passengers disembarking from ships, the police had to become even more involved in ensuring that drugs did not sneak into or out of the country. By 1920, the general practice was for customs officials either to try to delay passengers they suspected of having drugs or, in some cases, to follow suspects from the docks and notify the police of their whereabouts. Opium and cocaine cases thus came to be seen as the almost exclusive province of law enforcement by the time DORA 40B expired in 1920. As the Board of Trade explained in a November 1919 memorandum, cocaine and opium were now to be treated not as medicines or ordinary imports but more like firearms and other dangerous materials that posed a threat to public safety. Consequently, the task of drug law enforcement had become a police matter, both in theory and in practice.[30]

Shortly after DORA 40B came into effect, authorities were pleased with the regulation. In its first ten months, Liverpool police made 133 opium-related arrests, and in tandem with the power to deport aliens involved in opium smuggling, the new regulation enabled the authorities to put many traffickers out of business. By the summer of 1917, law enforcement officials in Liverpool claimed that opium smuggling in their jurisdiction had "almost ceased" and that shipping companies were satisfied with the progress in the campaign against opium smugglers. Elsewhere, however, investigations into violations of DORA 40B uncovered little of the illicit dealing or smuggling networks that the regulation had targeted. The Home Office concluded that most infractions of the regulation were of a "technical and minor" nature and resulted more from the administrative oversights of careless pharmacists than from the schemings of traffickers. Shortly after the regulation took effect, police reported that it had been effective in making doctors and pharmacists more prudent in their dispensing of drugs, and by April 1917, the Home Office would triumphantly conclude that there were "exceedingly few" recreational users of opium or cocaine on British soil. It was unclear, however, whether the lack of convictions under DORA 40B resulted from the salutary fear that the regulation instilled in would-be illicit dealers or from overestimates of underground trafficking before the regulation took effect. Nonetheless, law enforcement remained supportive of the new regulation and requested that the provisions of DORA 40B become law during peacetime.[31]

The Dangerous Drugs Act

Although DORA 40B gave authorities the tools to suppress opium trafficking and smuggling in Britain, it soon became clear that more needed to be done. In November 1917, news that a U.S. trafficker had used Britain as a launching point to smuggle four tons of opium to Shanghai revealed inadequacies in the existing control regime.[32] More striking, however, were revelations that emerged from the investigation into the November 1918 overdose death of well-known actress Billie Carleton.

The Carleton inquest revealed that, despite the Home Office's optimistic assertions concerning the effectiveness of DORA 40B, opium and cocaine were still available from many clandestine sources. Carleton, police found, had been using opium, morphine, and cocaine regularly for more than three years. She procured morphine from her doctor and got the other drugs through illicit channels—opium at a den frequented by London actors and singers, and cocaine from her costume designer, Reggie de Veulle. Subsequent investigations uncovered that de Veulle regularly got cocaine from a London pharmacist and opium from the owners of a Chinese knick-knack shop in East London, and that he hosted opium-smoking parties in his home. The prosecution of de Veulle and his associates was successful. A coroner's jury found de Veulle guilty of manslaughter for having provided Carleton with the narcotics that killed her. The shop owners who supplied him with opium were prosecuted (one was imprisoned, the other fined). And police identified a London pharmacist who had been feeding cocaine and heroin to the black market.[33]

Although a combination of existing laws and thorough investigation was able to break up the network that supplied Carleton, authorities saw the case as evidence that DORA 40B had not been "altogether successful" in controlling the flow of opium and cocaine. That such an elaborate web of dealers could function under DORA 40B spoke to the shortcomings of the regulatory regime. "It is clear from this and other cases that it is still possible to procure these drugs," the Home Office admitted, confessing that it did not have any idea "how the supply can be stopped except by making the penalties heavier." Newspapers, the *Daily Mail* and *Daily Express* in particular, agreed, noting that the revelations in the Carleton case "emphasized the urgent need of further drastic action to deal with illicit trafficking in poisonous drugs in London and elsewhere." Soon, others in the media picked up on the theme, publishing sensational stories on a possible drug epidemic seizing all of Britain. Recreational opium use, once a marginal pastime, now suddenly became "widespread" and "a menace to all classes of society," according to the *Daily Express*. The *Evening News* came to the gloomy conclusion that the country was in the grip of a "drug peril" that was attacking "more drug victims than ever." The *Times* even went as far as to term narcotics the newest "national vice." Despite DORA 40B, nightspots, doctors, pharmacists, and street dealers caught up in gambling and prostitution, claimed the press, all sold opiates and cocaine to anyone who was willing to pay the price.[34]

The solution, according to many in the media, was to pass a new law that would ensure better surveillance and prosecution of illicit narcotics dealers. "The ease with which victims of the drug habit and the traffickers in these poisons are able to evade the restrictions . . . has been proved in recent instances," the *Daily Express* wrote a few weeks after details of the Carleton case went public. "Additional precautions under the Defence of the Realm Act have not put an end to the trade, and it is obvious that drastic strengthening of the existing law is necessary." The main loophole in existing controls, the newspaper argued, was that police did not have the time or energy to give more than a quick glance to check that pharmacists' books were in order. The solution, they suggested, would be to hire full-time employees whose sole responsibility would be to inspect pharmacies. The *Daily Mail* agreed that better surveillance and more latitude to search and prosecute was the key; "the police are alert," it concluded, but "the difficulties of taking effective action" were too great for law enforcement to overcome. According to some press articles, stiffer penalties to deter would-be narcotics dealers and users would provide the best solution. The *Daily Express*, for example, argued that no pity should be given to the individuals caught up in the illicit consumption of or commerce in narcotics. "The drug-taker should no longer be looked upon as a weak-minded fool, but he should be dealt with as a criminal," the newspaper charged, while the "drug vendor, the vile instrument of this debauchery, should be dealt with as a felon."[35]

Intrigued by the newspapers' salacious details of the illicit drug trade and their suggestions on what to do about it, the authorities turned to the press for information and suggestions on how to address the narcotics problem. In particular, the police were puzzled that many details of the street drug trade printed by the papers had not come up in their own investigations. In January 1919, they contacted the editors of the *Daily Mail*, *Daily Express*, and *Weekly Dispatch* to find out their sources of information. After following up the leads given by journalists, however, the police quickly learned that the press was not privy to any inside information. To the contrary, much of what the newspapers had printed may have been fabricated. "The articles which have lately appeared are merely journalistic enterprise," the police concluded. "Possibly in the desire to add a little sensationalism to the articles . . . the imagination has been allowed to play a part in the preparation of the news." After closer investigation, the Home Office found that many of the allegations—that drugs were openly available in nightclubs, that certain doctors and pharmacists were dealing illicitly, and that drug taking was on the increase—were probably made by a single journalist with a hyperactive imagination and an aversion to fact checking.[36]

Nonetheless, the authorities saw the press coverage following the Carleton case as proof that more restrictive narcotics control legislation would be politically palatable. Officials sensed that the case provided both the impetus and the justification for a new law to further stem the availability of opiates and cocaine. The British public, at least according to the press, would favor more drastic state action against the narcotics trade. "It is now that the public is in the temper to demand that the police shall act at once and raid and shut the clubs wherein opium, cocaine . . . etc. have been sold for profit," the *Weekly Dispatch* claimed in December 1918. The Home Office soon picked up on the *Dispatch*'s cue. "In view of the undoubted growth of the drug habit, and of the strong public sentiment that the traffic should be brought under stricter control," the Director of Criminal Investigation wrote to the Home Office in February 1919, "the time seems to have come for enacting permanent legislation to take the place of Defence of the Realm Regulation 40B." The "strong body of public opinion in favour of permanent legislation" that appeared in the press, he reasoned, promised to give political traction to new and more comprehensive control initiatives. Taking advantage of the new prominence of narcotics in the pages of Britain's newspapers, Home Office staff made a point to collect newspaper clippings in the two months following the Carleton case. Even if the reports did not provide any new information, they were, as one anonymous Home Office official phrased it, "likely to be useful in preparing the public mind for possible legislation."[37]

Given that, in practice, drug control had become a police enterprise, the Home Office envisioned a new piece of legislation that would simply expand surveillance of narcotics and tighten controls over opiates and cocaine. The hegemony of law enforcement in drug regulation, however, was not uncontested. In particular, officials in the newly formed Ministry of Health also wanted to participate in drafting and implementing the legislation. To assuage these concerns, the Home Office agreed to work in conjunction with the Ministry of Health in writing the new law and formally agreed to rely on "interdepartmental consultation" when it came time to work out the specifics of implementation.[38]

Nonetheless, many medical professionals complained that the new restrictions created by the proposed legislation were far too tight. In particular, doctors and pharmacists resented provisions that required excessive bookkeeping and placed a ban on mail and phone orders for narcotics. Doctors maintained that such rules would "render the daily work of medical practitioners almost intolerable and interfere most gravely with their powers of relieving . . . acute illness," while also making the price of many medicines skyrocket. Pharmacists voiced even stronger objections, cursing the Home Office's proposed regulations as a "singular admixture of bureaucratic candour and bureaucratic ignorance" that ignored input from the professionals who would be most affected by the new law. In particular, they protested against a provision that proposed limiting the manufacture, sale, and distribution of substances classified as "Dangerous Drugs" to a handful of individuals licensed by the Home Office, instead of allowing all certified pharmacists to produce and sell them. W. J. Uglow Woolcock, a former secretary of the Pharmaceutical Society and a Liberal member of Parliament, complained before the House of Commons that such a provision would effectively supersede the Pharmacy Act, which had given permission to all licensed pharmacists to keep an open shop for drug sales. Given that opium was a common ingredient in medical preparations, he warned, pharmacists could see their business devastated if they were no longer allowed to sell it. Consequently, a chemist denied a Home Office license could, in effect, have "his business taken away" by the authorities, even though he was officially trained and licensed to practice the pharmaceutical profession. Woolcock also feared that if given too much power, law enforcement officials could withdraw pharmacists' licenses for "trifling" violations of the regulations, leaving members of the Pharmaceutical Society at the mercy of potentially arbitrary Home Office administrators.[39]

Others in Parliament feared the economic consequences of overly tight controls, arguing that stringent regulations could effectively "corner the market" in opiates for the few businesses licensed by the Home Office, inadvertently facilitating the creation of a price-gouging pharmaceutical cartel. Thus, although economic concerns had initiated opiate controls in Britain, they also gave good reason to hesitate before making the controls stricter. Under pressure from supporters of the medical and pharmaceutical professions in Parliament, the Home Office was compelled to make some changes to the proposed legislation, simplifying the recordkeeping requirements, exempting some popular medical preparations from the regulations, and removing some of the restrictions that would have curtailed doctors' ability to prescribe and purchase some drugs.[40]

The new legislation, passed as the Dangerous Drugs Act, 1920, was ultimately approved by Parliament and announced in August 1920. In spite of the concessions to medical and pharmaceutical interests, the new law entrenched the primary place of law enforcement in the administration and enforcement of Britain's narcotics control regime. Like DORA 40B, the Dangerous Drugs Act strictly regulated the manufacture, preparation, possession, and commerce of opiates and cocaine, and stipulated that any attempts to violate the regulations would be deemed an offense. It also tightened controls on narcotics imports, mandating that raw opium, its derivatives, and cocaine could be imported only with a government license, and only through the ports of London, Liverpool, and Southampton. Opium exports were subject to special restrictions to ensure that Britain would no longer serve as a global warehouse or transshipment point for smugglers. As under DORA 40B, the import and export of smoking opium was banned, as were the manufacture, sale, possession, and use of the drug. The new law also extended particularly tight restrictions to raw opium exports, dictating that shipments from Britain needed to comply with the laws of the destination country; if a nation had regulations banning or restricting opium, it was incumbent on the exporter to prove that the shipments would not enter the black market of the importing country. Production, sale, and distribution of morphine, heroin, and cocaine also became more tightly regulated, limited to authorized individuals specially licensed by the government. Only eleven firms were given permission to manufacture opiate medicines. For distribution to individual consumers, the controls also got tighter. Pharmacists were still allowed to sell these drugs, but medical practitioners were now subject to restrictions on prescribing. To ensure that the new drugs act would be effectively enforced, the government gave police some powers similar to those they had enjoyed during the war, such as the right to search, by force if necessary, any individuals or premises suspected of violating the drug laws of Britain or any other country.[41]

The most striking change from DORA 40B came in the penalties for offenses against the Dangerous Drugs Act. Violations of DORA 40B were punished with fines of up to a hundred pounds and six months imprisonment;

the Dangerous Drugs Act multiplied the penalties, imposing fines of up to a thousand pounds, imprisonment for up to ten years, or both. If a corporation broke the law, its chairman and directors would face criminal penalties, unless they could prove they were unaware of the corporation's illicit dealings. From the street corner to the boardroom, therefore, involvement in the drug trafficking business now became an exceedingly risky proposition. Despite the large amounts of money to be made on the black market, authorities hoped that the Dangerous Drugs Act's stiff penalties, coupled with the broad powers it gave to police, would deter would-be smugglers and illicit dealers.[42]

In the first years the new law was in operation, the Home Office was not reluctant to use it. Between 1921 and 1924, it opened 835 cases for violations of the Dangerous Drugs Act: 482 involved opium, 291 involved cocaine, and a handful involved morphine, heroin, and other substances. The punishments meted out to offenders, however, were relatively mild compared with the allowable maximum sentences. A large number of cases were either dropped or resulted in fines of less than ten pounds, and only 222 led to prison sentences. The most severe treatment was reserved for immigrants and foreigners, seventy-seven of whom were recommended for deportation after serving their time or paying their fine. A few years after the Dangerous Drugs Act took effect, it seemed that the law enforcement was working as a deterrent: the number of prosecutions for drug law violations dropped from 277 in 1923 to just 84 in 1924.[43]

The number of cases involving the Dangerous Drugs Act remained quite low until World War II—fewer than a hundred cases each year until 1941, with a low of just thirty-nine in the entire country in 1937. Beginning in the early to middle 1920s, doctors who specialized in treating addiction started to see fewer addicts, noting, as one practitioner observed, that "the obstacles now placed in the way of procuring the drugs" were making opiate habits "much less common than before the War." The Home Office was pleased that the policy was serving as an effective deterrent, coming to the conclusion, in 1925, "that most of this reduction can fairly be taken [as being] to the credit of the effective operation of the Acts." While stricter enforcement played a part in the drop in violations, the Home Office also noted that increased care by pharmacists—who began to follow the rules on drug sales and documentation more closely once they faced the prospect of heavy fines for noncompliance—also played a role. Furthermore, the amount of synthetic drugs manufactured in Britain dropped drastically; with the issuance of licenses to just a limited number of factories, the amount of morphine and heroin produced in the country decreased by more than eighty percent from 1920 to 1921.[44]

With fewer opiates on British territory, and in fewer hands, the task of surveillance became much easier for the authorities. Nonetheless, unsanctioned drug transactions and use may have continued, since enforcement was spotty at best. "The bulk of the prosecutions were undertaken by about a dozen forces," the Home Office observed in its analysis of Dangerous Drugs cases, leading to speculation that drugs may have been circulating free of government surveillance in areas where the police were less vigilant. Historical research has indeed confirmed these suspicions. Oral histories of pharmacists who ran shops in the 1920s and 1930s reveal that in some areas, opium preparations, and even opium poppy capsules, continued to be available over the counter throughout the interwar period. This was a problem of implementation rather than legislation, however, and by the mid-1920s, authorities believed that the flow of narcotics on British territory was well on its way to being effectively controlled. "The action taken by the police to enforce the law has been successful to a very large extent," the British delegation proudly reported to the League of Nations in 1923, claiming that outside a few enclaves of Chinese opium smokers in port cities, the practice of dealing and indulging in these drugs had effectively come to an end.[45]

In the 1910s and 1920s, anxieties about public health and morality, even though manifest in the writings of journalists and medical researchers worried about addiction, played a secondary role in the development of drug control initiatives that were actually put into place in Britain. Concern about the nation's commercial well-being—which seemed imperiled by potential sanctions on international trade in the 1910s—proved to be the deciding factor that moved the government to initiate action against the unregulated opiate trade on British soil. Opium prepared for smoking was the most tightly restricted drug under both DORA 40B and the Dangerous Drugs Act, even though authorities believed that the use of the substance in Britain was limited and well under control. Where opium prepared for smoking was a major problem, however, was in the commercial sphere, as it lay at the heart of the Britain-based smuggling operations that imperiled international trade. To address the issue, both DORA 40B and the Dangerous Drugs Act particularly focused on controlling the export from Britain of opium prepared for smoking. Thus, while early British narcotics control policies certainly touched on other substances that British citizens were using (e.g., morphine, cocaine), their central target was a drug (opium prepared for smoking) that was being smuggled much more than it was being consumed. Repressing the international drug traffic, more than limiting domestic drug use, was ultimately the principal aim of Britain's early drug control regime.

This was not the case in all countries. The path toward national narcotics control would go in a different direction when the drug problem was framed not as an economic threat but as a sociopolitical one.

France in the Age of National Opiate Control

The Move toward Opiate Control

Until the twentieth century, France, like Britain, controlled opiates with pharmaceutical regulations. Opium and its derivatives were subject to poisonous substances laws that dated back to 1680, and more stringent controls that came into effect in the 1840s. Initiated because of concern about the injudicious use of chemicals following some high-profile poisoning cases, a series of initiatives—an 1845 decree, an 1846 royal ordinance, and a subsequent decree in 1850—were designed to limit the availability to the general public of potentially dangerous substances. These regulations stipulated that individuals who sold poisonous substances (including opiates) needed to register with municipal authorities and that wholesalers could sell the substances only to pharmacists or individuals who held a government license. Vendors had to keep registers tracking their poisons purchases and sales, the amounts involved, and the names, addresses, and professions of all parties involved in the transactions. Only licensed pharmacists could retail medicines containing poisons, and only to individuals who presented a written prescription from a doctor, surgeon, licensed health officer, or veterinarian. The regulations also specified that prescriptions should be signed, dated, and written in print to avoid confusion over penmanship and should specify the prescribed dosage and mode of administration. Pharmacists were required to keep the poisons registers (tracking the comings and goings of controlled substances) for twenty years, and their records were subject to inspection at any time. Furthermore, poisons had to be kept under lock and key and stored in a separate cabinet to prevent accidental contamination of nonpoisonous preparations. The penalties for disregarding these regulations were severe: violators could be fined up to three thousand francs and face up to two months in prison. Thus, unlike in Britain, where opiates had been openly available, the drugs (at least in theory) were tightly controlled in late-nineteenth-century France.[46]

Notwithstanding the thoroughness of the law on paper, the regulations had little effect on the availability of opiates in everyday life. In many regions, pharmacists did not keep the required registers, and when violators were caught, the scope and scale of their offenses revealed how ineffective the regulations were. At times, pharmacists sold enough morphine to individual customers to satiate even the heaviest of drug habits for years, and many wholesalers ignored the stipulations of the law and sold directly to the general public. Furthermore, authorities were powerless to stop individuals ordering drugs through the mail from unregulated foreign manufacturers and pharmacists. Thus, even with a thorough regulatory apparatus and severe fines for violations, poor surveillance and shoddy enforcement rendered the poisonous substances laws effectively moot. According to many French medical practitioners concerned with opiate addiction, the government had done an admirable job in writing regulations but had failed when it came to enforcing them. "The laws exist in France," researcher Georges Pichon complained in his 1889 study of morphine addiction. "*It is their application* that is the problem."[47]

Frustrated when calls for more vigilant enforcement fell on deaf ears, doctors concerned about the open availability of opiates considered proposing new legislation to impose tighter controls. In July 1891, for example, the Comité consultative d'hygiène publique debated drafting a proposal for the Chamber of Deputies that would have prohibited the renewal of morphine prescriptions. Noting, however, that most merchants and addicts took little notice of pharmacy regulations anyway, the committee decided that adding another provision to an already ignored law would do little good, and dropped the idea. "The answer is very simple," it reasoned, advising that a combination of surveillance and "severe punishment" of violations would solve the problem. In 1895, the Minister of the Interior took notice of the issue and solicited the opinion of Paul Brouardel, a professor at the Paris Faculty of Medicine, for input on what could be done to check the spread of opiate addiction. Brouardel suggested banning renewals of opiate prescriptions and imposing harsh penalties on individuals who tried to use fake prescriptions and on wholesalers who distributed drugs directly to customers. Brouardel recognized, however, that the task of primary importance was applying the regulations more strictly. Short of putting opiates under a completely different control regime, the best course of action seemed to lie not in any new legislative initiatives but in the more rigorous enforcement of existing measures.[48]

Little came of the medical community's efforts to cut down on the availability of opiates, but the concerns of another institution—the military—resulted in more direct action. When military officials learned that levels of recreational opium use in the navy and colonial army were on the rise, they began exploring ways to address the problem. Regulations forbidding opium smoking by servicemen held the promise of dissuading some from engaging in the habit, but Minister of the Navy Gaston Thomson remained concerned that the wide availability of opiates would compromise any military anti-drug campaign. Unlike doctors, who blamed wholesalers and pharmacists for distributing drugs too liberally, military authorities suspected that other individuals—smugglers, prostitutes, and even soldiers—were supplying opium dens in French port cities. In 1906, Thomson initiated a study of the problem, ordering maritime police and civil authorities in southern France to explore how to keep opium away from military men.[49]

The investigations revealed that opium was widely available in Toulon. It was used by most demimondaines in the city, particularly when entertaining clients from the navy and colonial army. As the inquiry progressed, authorities found that the drug was more openly available than they had thought. Within weeks, they discovered that two individuals, former colonial officer JP and demimondaine BS, ran opium dens that were open to the general public, and that they were supplied by a Toulon shop owner, JG. Even more disturbing for the military was the discovery that the dens were not just centers of drug use but potential threats to national security. Police learned that at JP's den, a prostitute would befriend young soldiers who were under the influence of opium and try to seduce them into divulging information "pertaining to national defense." When the prostitute was arrested for espionage, the case reconfirmed the links between opium and betrayal that had become prevalent in late-nineteenth-century opium literature and would later be reinforced by the Ullmo Affair. Opium, it seemed, was becoming more than just another on the laundry list of unhealthy vices indulged in by soldiers; as a drug that France's enemies were using to lure young men into betrayal, it posed a fundamental threat to the ideological health and discipline of the armed forces.[50]

Once made aware of the extent and potential consequences of the opium habit in port cities, the Minister of the Interior ordered local law enforcement to rigorously enforce poisonous substances regulations throughout southern France, and opened investigations at the homes of JP, BS, and JG. In their searches, authorities found large amounts of opium, as well as a hefty collection of opium-smoking paraphernalia. Though without evidence, other than hearsay, that the three individuals had participated in opium transactions, the authorities deemed the discoveries adequate proof that all three had participated in the underground drug trade. Subsequent investigations backed up these suspicions, particularly when detectives found handwritten letters from JG saying that he sold more than 350 kilograms of opium per year, and witnesses confirmed they had paid to smoke in JP's den. That December, guilty verdicts were returned against JP, who had to pay a fine of two hundred francs, and JG, who was penalized a thousand francs.[51]

Even though these investigations led to successful prosecutions under the existing legislation, authorities feared that the JP-JG ring was just one of many such clandestine operations running throughout the region. Police investigations showed that JG's main supplier was, as authorities had suspected, not a pharmacist or pharmaceutical wholesaler but a sailor who smuggled opium from Indochina into Marseille.[52] The involvement of smugglers in the domestic opium trade promised to greatly complicate matters as authorities tried to restrict the availability of the drug for recreational use. Whereas pharmacists could be monitored with the register provisions of the poisonous substances acts, individual smugglers kept no records that could provide documentary proof of illicit transactions. Moreover, sailors constantly moved from port to port, making it difficult for detectives to investigate their smuggling activities in any detail before the seamen left town. Given these factors, existing pharmacy legislation and enforcement measures were poorly equipped to attack the underground opium trade that was flourishing in southern France.

To make matters worse, other investigations revealed that the problem was not confined to southern parts of the country. In January 1907, police found that soldiers stationed in Brest regularly smoked opium with prostitutes, who procured it with the complicity of local pharmacists. In the fall of that year, the Ullmo Affair brought another example of opium's availability into clear relief. In December, police followed an anonymous tip to the home of two Marseille shopkeepers, where they found large stores of opium. The following spring, another tip led police to a Parisian who sold opium to military personnel stationed in the capital. Some of the offenders in these cases were prosecuted and fined, but the apparent prevalence of underground opium dealing across France caused alarm in the high command of the armed forces.

Throughout the country, warned military officials in a note to the Ministry of Justice, opium was "spreading throughout the army" much more quickly than was first believed. As a growing threat to the health, discipline, and dedication of the armed forces, opium was developing the potential to become a security menace of the first degree.[53]

By the summer of 1907, Thomson had become convinced that existing measures were not doing enough to keep opium out of military hands. In September, he wrote to the Ministry of the Interior asking that steps be taken to curb the opium habit in the military. The best way to do this, he suggested, was to regulate not only the sale and purchase of opium but also its use. Officials in the Ministry of Justice, however, were not willing to go as far as to ban the practice of opium smoking, arguing that the government would overstep its bounds if it tried to regulate an individual behavior such as drug use with an executive decree. It could, however, target the practice by placing further restrictions on opium commerce, stipulating that the drug could be sold only for medical use. Such restrictions, Justice officials reasoned, were legally justifiable because they did not prohibit the act of smoking opium, but only regulated commercial exchanges involving the substance. Furthermore, restrictions on sales could accomplish what Interior officials wanted by creating an opening for the prosecution of both sellers and recreational users. By placing conditions on the purpose for which opium could be sold, such a provision would put the onus on pharmacists and wholesalers to ensure that the opium they doled out did not go toward recreational use. By extension, such a provision could also make opium smokers themselves liable to prosecution, since any opium used other than as a medical treatment would become, by definition, "illicitly purchased." By leaving the term "medical purposes" undefined, the regulation would put the burden of proof on individuals found using the drug to make the case that they were using it out of medical necessity, not simply for pleasure. Adding this restriction to opium sales could bring both the individuals who supplied opium dens and those who smoked the drug within the reach of the law.[54]

Tighter regulations governing commercial exchanges of opium, however, were not enough to tackle the problem effectively. Given how easy it was to deal the drug illicitly, and the potential profits involved, even the most severe controls over opium exchanges would not be enough to check the underground flow of the drug through France's port cities. "The clandestine importation of opium will always be tempting to smugglers," officials from the Ministry of the Interior argued, "thanks to the product's light weight and high price." Thus, beyond simply putting new regulations on the exchange of opiates, law enforcement also recommended expanding controls to the arenas where most recreational opium use took place: opium dens. The way to "stop the problem," the Ministry of the Interior explained, was to "directly target the managers" of the places where individuals gathered to smoke the drug. Existing laws allowed the prosecution of individuals in charge of opium dens only if they were caught in the act of selling the drug, and law enforcement had to show that den managers were selling the drug for profit. This was difficult to prove, and sometimes den managers did not charge for the drug, but instead collected money for entrance into the den and provided opium free of charge. With this indirect system of opium dealing, den managers had devised a way to protect themselves from prosecution under the poisonous substances laws. To close this loophole, the Ministry of the Interior suggested a provision that would take aim at the "people who tolerate the breaking of the rules concerning this matter in their own homes"—in other words, making it illegal to run a private opium den.[55]

To institute such a regulation on its own, however, would have been problematic, for it was not possible to create an offense for facilitating or tolerating a behavior (smoking opium) not yet legally proscribed. "Can the den manager be targeted," Ministry of Justice officials wondered, "if, without transferring or harboring opium, he manages an establishment that is open to the public? And can this be done without a new legal disposition?" The answer would lie, once again, in the restrictions on opium sales. If it was made illegal to purchase the drug for recreational use, the government reasoned, individuals who facilitated nonmedical use could come within the reach of the law; given that opium dens were used to put the drug to purposes other than those for which it could be legally sold under the proposed regulations (as medical treatment), the men and women who ran such establishments could be considered parties to a crime. Opium den managers could then be prosecuted as accessories to a law-breaking activity. By extending the restriction of opium sales to medical usage, the Ministry of the Interior reasoned, authorities could make it an offense "to encourage the illicit use of opium, be it by consenting for the use of a premises for the smoking of opium, or be it by any other means." Such a provision would be particularly useful in targeting not just the individuals who ran and frequented opium dens but those who supplied them with drugs and smoking paraphernalia. "Anybody who, in a public space or a space that is accessible to the public, encourages opium use by *helping* or assisting in procuring the means (pipes and other objects used by smokers)" to consume opium could be punished as an accomplice in the committal of an offense. Thus, by extending the restrictions on sales to other activities that could be considered "preparatory" offenses, the new regulation—without taking the step of making opium smoking illegal—could place significant practical and legal roadblocks in the way of smokers when they tried to procure the drug or attend the dens where they usually consumed it. Striking at users, dealers, and den managers, such restrictions on sale and "encouragement" held the promise of effectively putting an end to recreational opium use in France.[56]

The Minister of the Interior announced the new regulations in a decree issued on October 3, 1908. The first provisions of the new rules instituted a tighter system of control over opium once it entered French territory; importers were required to get receipts from customs officials indicating the amount of the drug received and the name and address of the consignee. Within three months, this document had to be forwarded to the municipal authorities in the area to which the drug was being sent, thus ensuring that local officials could track all of the opium in their jurisdictions. Only wholesalers, industrialists, chemists, and pharmacists were allowed to acquire the drug from importers, and only for medical research or usage. Opium purchases, as had been the case under the poisonous substances regulations, were to be recorded on a register to track the comings and goings of the drug, and sales could be made to the general public only with a medical prescription. Thus, from the time opium entered through French ports to the moment it was consumed by the patient, all transfers of the drug had to be officially documented. The decree specified that opium could be transferred only for medical use, thus banning its sale for recreational purposes, as well as forbidding participation in other activities (running an opium den, selling smoking paraphernalia) that facilitated recreational opium smoking. In addition, to shut down opium dens, the new law included a provision that made the transfer of opium, "even for free," illegal if it was not going to be used medically. Thus the actions of den managers who did not charge customers for the drug now came within the reach of the law.[57]

The decree also included a critical provision that made it an offense to "encourage the possession or illegal use of opium, by allowing the use of a premises or any other means." The vagueness of this phrase opened a potentially wide legal door for the authorities. Since "encouragement" was not legally defined, it could encompass a variety of activities—from telling individuals they should try opium to peddling the drug on the street. Making "encouragement" an offense, the new decree gave police carte blanche to pursue cases against anyone they suspected was involved in the proliferation of opiate use, even if not caught in possession of the drug or in the act of selling it.[58]

In spite of these changes, the penalties for violating the new decree—a fine of up to three thousand francs and up to two months in prison—were identical to those for offenses under the poisonous substances acts, and remained insufficient in the eyes of many law enforcement officials. Nonetheless, authorities considered the decree a potentially valuable tool in their work to check opium use, one that could fill important gaps that had existed under the pharmacy laws. Through "exact and rigorous" enforcement, along with widespread surveillance, searches, and seizures, the Ministry of Justice envisioned a new control regime that could soon make the opium dens of France things of the past. Yet, as the following years would show, the decree alone would not be enough to control the flow of opium on French soil.[59]

The First Battles of the French Drug War

In the eight years that the 1908 decree was in effect, it had mixed results. Enforcement of the new opium regulations was generally limited to the jurisdictions where the problem was known to be most widespread—the port cities of southern France, Brest, and Paris. From the decree's coming into effect through the end of 1913, law enforcement investigated 124 cases involving opium violations of the pharmacy laws and the 1908 decree, charging 182 individuals with crimes and finding 136 of them guilty. Authorities observed that the "public" opium dens, ones that were open to the general public but charged an entrance fee, became exceedingly rare in the five years after the new regulations were issued; by the spring of 1913, the Aix prosecutor would proudly assert that such dens had all but "disappeared." Nonetheless, opium remained available through a wide variety of illicit channels. Seizures of more than 150 kilograms of illicit opium in the first six months the decree was in force revealed the breadth and scope of the problem, especially since police intercepted just a small fraction of the drugs that reached the black market. Even more worrying for the authorities was that the drug habit remained prevalent in military circles—the very population the decree was designed to keep away from opium. Despite increased vigilance by the military to curb use in its own ranks, investigations in port cities revealed that the practice remained endemic in the armed forces; officers met in secret to indulge in the drug, and demimondaines "readily" invited soldiers to smoke opium with them. As a detective investigating Toulon opium dens grimly summed up in 1909, "out of every ten [opium] smokers, you can be sure that at least six of them are soldiers."[60]

Since the 1908 decree regulated both the sale of opium and, indirectly, its use, authorities faced two distinct tasks when enforcing the new regulations. First, given the focus on limiting commerce in the drug, the pharmacists and smuggling rings that supplied individual smokers and the managers of opium dens were the principal targets of law enforcement investigations. The pharmacists and wholesalers feeding the illicit market were relatively easy to identify, as their registers provided documentary proof of transactions, with evidence of large sales and the names of customers who purchased with suspicious regularity. Yet the cases against pharmacists rarely uncovered any significant suppliers of the underground opium market; most of the pharmacists prosecuted for violations of the 1908 decree were guilty of minor bookkeeping errors, though a few distributors for the black market were discovered when the authorities checked opium registers. More important, however, these cases instilled a salutary fear throughout the pharmaceutical profession. By the middle of 1909, most pharmacists took care to keep their registers in order and to refuse the sale of opium to customers who did not present a prescription.[61]

Shutting down the supply of opium from shops, however, opened a Pandora's box for the authorities, as the underground market for the drug grew to fill the gap left when pharmacists began obeying the law more strictly. Herein lay the second task: tracking and gathering evidence against street dealers and the unknown suppliers for dens—work that was much more difficult than monitoring the goings-on in shops that sold pharmaceutical preparations. Pharmacies were stationary, had set hours, and were open to the general public, but street dealing involved itinerants, took place at all hours of the night, and was carried out in a secretive manner designed to avoid detection. Thus, although driving supplies of recreational opium further underground promised to make it more difficult for users to procure the drug, it had the unintended consequence of accelerating the development of a black market—one that would be much more challenging to track.[62]

Given the difficulties inherent in suppressing smuggling and street dealing, French law enforcement turned its efforts toward the detection and prosecution of individuals who ran or frequented opium dens. Compared with the ever-shifting world of suppliers, the circle of opium users who gathered at dens was a relatively sedentary population, and one that met regularly in set locations. Thus, instead of chasing directly after the heads of opium-smuggling syndicates, the French approach was to work up the supply chain, prosecuting and interrogating users, den managers, and street dealers in order to identify and eventually arrest the individuals in charge of more large-scale operations.

Penetrating an opium den was a time-consuming and difficult task, requiring detailed surveillance and extensive undercover work that was beyond the normal purview of most local police. Outside Paris, the law enforcement officials most up to the task were members of the Sûreté générale (SG), a nationally centralized group of detectives that specialized in assisting local law enforcement in investigating espionage and organized crime. In particular, veterans of the Brigades des courses et des jeux, a branch of the SG that focused on underground gambling rings, were well-suited for the job of uncovering opium dens. In the repression of illegal gaming, these agents had proved more effective than local detectives, for two main reasons. First, unlike municipal law enforcement officials who were well-known to the overseers of secret gambling clubs, detectives could arrive from Paris incognito, unrecognized by locals and better able to penetrate criminal syndicates. Second, local police who grew familiar with the bosses of gambling rings often took bribes or were bullied into turning a blind eye to illegal activities under their watch. The skills required to penetrate opium dens, reasoned the authorities, were similar to those that SG detectives had developed in their investigations of gambling rings, particularly the abilities to search for secret locations and work undercover. The necessity of sending special agents to investigate the illicit opium trade became abundantly clear to detectives when they observed the inability of local authorities, outside Paris, to uncover opium dens. "In the circles where people smoke opium, they never think that the police could intervene," an SG detective on assignment in Lorient observed in May 1909. Local police, he concluded from his initial investigations in several cities, were either lazy or incapable and "never took action" even when opium dens were operating right under their noses.[63]

Usually working alone, SG detectives were sent from Paris at the request of local authorities, or arrived on their own initiative if investigations gave reason to suspect there was an underground drug ring in a given city. Central authorities in Paris assisted the investigators, giving them a per diem and a separate fund to pay off informants. When they arrived in a city, detectives generally began their work by consulting with local police for leads, then going undercover—posing as out-of-town businessmen, adopting pseudonyms, and searching the town for opium dens. They often pretended to be opium users looking for drugs, and they gathered tips from bar patrons and demimondaines to lead them to dealers or dens. The agents then compiled names, addresses, and as much detail as possible about the individuals or locales rumored to be involved in the underground opium trade, and used this information to set the agenda for subsequent surveillance. At times, the tips from locals amounted to little more than gossip, but sometimes they led to more significant revelations, such as the addresses of apartments where private opium dens were run, or the locations of opium stores and public dens in the backrooms of bars, cafes, and clandestine casinos. If a lead seemed promising, investigators staked out suspected dens and storage locations, tracking the comings, goings, and activities of individuals who frequented the premises, for up to a week. Sometimes they went a step further, arranging meetings with reputed opium dealers, and even attending opium dens and smoking a few pipes to gain the trust of the individuals they were investigating. Some detectives who smoked regularly as part of their stings became opium aficionados, at times even sneering at the poor quality of the drugs they discovered. One agent on a mission in Brest in 1913 told his superiors that, while undercover, he had recently "smoked . . . a disgusting opium that seemed more like wax than anything else." He even admitted, without shame, that he had indulged a little too much during one sting and fallen under the influence of the drug while on the job. Yet, despite the dangers inherent in such "participant observation," no detectives—at least among those whose notes and reports appear in the SG archives—ever admitted to blowing their cover. Smoking opium undercover, ironically, became an incidental perk of some investigators' jobs.[64]

Once undercover investigations were completed, it was the job of local law enforcement to use the information gathered by SG officials to collect evidence and prosecute users, den managers, and opium dealers. Yet, as in the undercover work that went into opium investigations, local police outside Paris were grossly ineffective. In 1909, for example, a detective sent to Lyon spent nearly a month investigating the local opium trade, but by the time local authorities took action—some three weeks after he filed his final report with them—the intelligence he had gathered was out of date and the detective's confidentiality had been compromised. Conflicts of interest also proved difficult when powerful individuals were involved; police did not take action when judges or one of their own were implicated in trafficking or smoking opium. Throughout their correspondence with their superiors in

Paris, detectives working undercover complained that the local police were at worst corrupt and at best incapable. It seemed, one detective quipped, that local law enforcement had "given their most incapable agents the task" of assisting in opium investigations. The Rennes prosecutor did not mince words in a 1913 report to the Minister of Justice, terming the local police "incapable, ignorant," and "unable to gather any useful information." Furthermore, the reticence of some judges to enforce the law did not help matters. Sometimes they refused to hear cases that implicated respected members of the community or army, no matter how compelling the evidence.[65]

Limited—and perhaps actively hindered—by the inactivity and bumbling of local authorities, many exhaustive SG investigations led to no convictions. This was particularly frustrating because reports from both the military and civil authorities in port cities pointed to the continuing prevalence of the opium habit in the armed forces, with some sailors and administrators even setting up private opium dens in their barracks and offices. Beyond failing to achieve their principal goal—to eliminate opium smoking from the military—these investigations also fell short in discovering who was responsible for feeding the illicit market. By 1913, the underground trade seemed to be booming, and dealers grew more aggressive, even making offers to sell as much as fifteen kilograms to passersby on the street. Without testimonies from convicted users, small-time dealers, and opium den managers, it was difficult to identify traffickers and build cases against them; consequently, the black market continued to flourish.[66]

Failed prosecutions frustrated not only the detectives who carried out the investigations but the higher-ups in law enforcement and the general public. In 1912, the prosecutor for the Paris region wrote a scathing memorandum to his subordinates, taking them to task about opium and morphine seemingly being available "without any difficulty" to all in the capital who wanted them. By late 1913, the perception that the authorities either were unable to stop or were complicit in the underground opium market spread to the populace and the press. One concerned citizen wrote to the Ministry of the Interior, charging that informants used their police protection to openly sell opium out of a Toulon café and that the authorities turned a blind eye to drug-dealing hotspots. *L'Eclaireur*, a Nice newspaper, lamented that despite intensive investigations, law enforcement still did not nab any dealers. *Le Matin* complained that "legions" of drug dealers walked the streets of Paris. And *Le Petit Provençal* charged that nearly all demimondaines in the Côte d'Azur were involved in drug trafficking.[67]

The overdose death of actress Pierrette Fleury at a Versailles estate in September 1913 led to speculation that country retreats around Paris were "drug villas" that the well-off used to indulge in recreational drug use, and that police were doing nothing about it. In 1914, Delphi Fabrice, an investigative journalist, went a step further, publishing a book claiming that the opium habit was spreading across all classes, ethnicities, and neighborhoods in the capital. Some journalists, most notably *Le Matin* editor E. Rouzier-Dorcières, went as far as to accuse law enforcement of "tacitly authorizing" opium dealing by providing protection for street dealers. On one occasion, Rouzier-Dorcières even marched into the headquarters of the Toulon police and challenged them to prove they were not involved in drug trafficking. The official response to such charges, especially those from Rouzier-Dorcières, was often defensive. In 1913, the controversy over his bombast reached a boiling point when a naval doctor, upset at the editor's willingness to besmirch the good name of military personnel, challenged him to a duel (it resulted in a draw).[68]

Despite the perception that their anti-drug efforts were failing, authorities were able to identify sufficient numbers of major smuggling operations to gain insight into the tricks of the underground drug trade by the beginning of World War I. In particular, investigations in southern France gave the authorities a good idea of how opium made its way into the hands of users across the Côte d'Azur. The center of the illicit traffic in southern France was Marseille, a port frequented by both military and commercial ships from Southeast Asia. The opium used by military men was sometimes smuggled by members of the armed forces themselves; officers stationed in the Far East could bring drugs home undetected, since customs rarely checked the baggage of uniformed servicemen.[69] More difficult, however, was the task of discovering civilian smugglers. Unlike the military command, civil authorities did not have the resources necessary to uncover the smuggling rings that operated in Marseille. To complicate matters further, the majority of the civilians engaged in smuggling were not French citizens but Asians and Middle Easterners, who were rarely on French territory for more than a few days at a time. Unlike the smugglers in the armed forces who distributed drugs by the kilogram, traffickers on commercial liners carried opium in such small amounts that it was very difficult to detect.

As authorities stepped up their searches of luggage on commercial vessels, smugglers adapted, hiding their stores in everyday items—hollowed-out umbrellas, canes, vases, tea sets, drums, or cigarettes. Consequently, though customs agents could search passengers' baggage, they had neither the man-power nor the perseverance to check every item brought to shore. Even transferring just a few ounces of opium at a time, given the prevalence of small-time smuggling, added up to a significant amount of drug that escaped control. "The quantities introduced this way are minimal," the Aix prosecutor explained in a report on the opium trade in his jurisdiction, "but for the crew who make this voyage regularly, this method . . . can bring a significant amount of opium to land." As stricter enforcement in southern cities made opium more scarce, the price of the drug went up, enabling traffickers to invest more in their smuggling schemes. By 1913, traffickers would send private motorboats into ports to unload large contraband shipments that smugglers threw overboard from ships, and they spent large sums of money bribing customs officials to ensure their cooperation. Once they came·to shore, drug shipments were divided up and distributed either directly to users or to middlemen—usually bellhops or waiters—who sold directly to opium dens. Other portions of the smuggled opium were distributed to bars and hotels throughout the city, for peddling to potential new customers. The lion's share of contraband opium, however, was purchased by traffickers who forwarded the drugs to Toulon, where demand was high and the market particularly lucrative.[70]

As Marseille-based operations grew in the years after the 1908 decree, opium trafficking became integrated into the worlds of organized crime and large-scale smuggling. By 1909, one detective in Marseille reported that sailors would smuggle opium every time they returned home, and that the drug was becoming a new cornerstone of the established underground market for other goods, such as tobacco, that were smuggled into the port. The increasing prevalence of career criminals in opium dealing reflected the drug's new place as a mainstay of the black market. Investigations also revealed that suspect foreigners—Germans in particular—were involved in Marseille's street dealing, which, from a national security standpoint, was particularly troubling for authorities. Revelations that a Toulon demimondaine involved with some German street dealers had been sending drawings of French warships to a German engineer seemed to confirm the connection between the opium traffic and espionage. As one undercover detective reported, the woman even quipped that she considered "Ullmo to be a brother" in spirit. Reports of such remarks, combined with the role that opium had assumed in the French political imagination, reinforced an official view of the drug trade as an enterprise that was not only distasteful and unlawful but potentially treasonous and anti-patriotic.[71]

Outside the southern port cities, the main method of distributing smuggled opium in France was through the mail. Often, importers-exporters procured the drugs from Marseille smugglers, using their legitimate businesses as fronts for trafficking. These dealers then sent drugs to shops throughout the country, particularly in Paris, that doubled as opium dens after hours. By 1914, such suppliers took care to thwart potential surveillance of mail customers' orders, always having them write to request "embroideries" or other innocuous-sounding code words for the drug. Sometimes medical students, who also had access to stores of opium, became involved in the underground trade, mailing drugs to prostitutes and demimondaines who dealt on the street. Such traffickers, however, were a minor nuisance to law enforcement when compared with syndicates that mixed trafficking-by-mail with traditional smuggling and dealing methods, beginning in late 1910. The most notorious of these gangs was the one led by a failed journalist, Louis Lardenois, and his wife, Camille Graver, who was a dancer at the Moulin Rouge. Instead of procuring supplies from southern France, Lardenois and Graver purchased opium in London and smuggled it into France either by mailing it or by carrying it with them on boat trips. By 1915, Lardenois was transporting more than forty-six thousand francs worth of opium to France each year, and his account books reveal that he enjoyed a profit margin exceeding thirteen hundred percent on his opium transactions.[72]

By 1911, the perception—both within and outside the halls of government —that the 1908 regime was ineffective led to calls for change. Many public figures, such as Ille-et-Vilaine Senator Léon Jenouvrier, maintained that the problem lay not in the existing law but in its implementation, and that there was plenty of blame to go around. "Normally there is just one specific ministry to blame," Jenouvrier told *Le Matin* in 1913, "but now I do not know which one to talk about, since there are five ministries who are doing nothing." Pointing a finger at the justice system that did not effectively prosecute cases, the police who did not investigate opium dens rigorously enough, the naval authorities who did not effectively suppress the habit within their own ranks, the customs officials who did not stop smuggling, and the colonial administration that allowed the drug to be grown in Indochina, Jenouvrier found the entire government guilty of allowing opium traffickers and users to continue their activities with little difficulty. The consequences of the spread of the opium habit were potentially disastrous, he argued, because of the drug's effects both on the physical health of users and on their spirit. "Opium is a moral poison, a social poison," Jenouvrier charged, one that could break down not only the physical body of the user, but also the social body of the nation.[73]

Commentators such as René Le Somptier agreed, forecasting that France would soon become "decrepit" if the drug habit continued to spread unabated. "Under the influence of this criminal alkaloid," he warned, Frenchmen would become increasingly apathetic, simply "closing their eyes and sleeping" instead of fulfilling their duties as citizens. Such an "infiltration of this murderous poison into our democracy," Le Somptier wrote, could effectively kill the Republic by sapping its political and moral will.[74]

Anxieties over the ideological consequences of the spreading opium habit became manifest through the increasingly common linkage between the drug and France's main enemy—Germany. Beyond the concern that the Germans could use the drug as an agent of treachery, Frenchmen became convinced that opiates were part of a covert German plan to poison the nation. Even before the outbreak of World War I, insinuations to this effect were common in discussions of the drug traffic. Such speculations were not completely outlandish. The German pharmaceutical companies Merck and Boehringer indeed manufactured a large proportion of the morphine and heroin that made their way onto French streets. Yet, given that the French colonies were also a major source of the opium on the black market, the argument could have been made that the French colonial administration as much as the German pharmaceutical industry was to blame for the prevalence of the problem.

Generally speaking, however, that semisynthetic and synthetic opiates were manufactured in Germany seemed compelling enough evidence for many people that Germans were conspiring to flood France with the drugs. *Le Matin* was particularly insistent on this point, continuously emphasizing that much of France's morphine came from Germany and that German scientists had invented the drug. Paul Sollier, director of the Boulogne-sur-Seine sanatorium, agreed that it was Germans, in particular, who were responsible for initiating and proliferating France's drug problem. "Germany inundates us . . . with new narcotic products," he asserted in 1912. "Dionine, paulopon . . . all successors of morphine, derivatives of opium, that are addictive," were all created in German laboratories and, he warned, could potentially come to France as new waves of a narcotic invasion.[75]

The combination of increased concern about German links to the spread of opiate use and the apparent powerlessness of the authorities to stop it led members of the legislature to take action. In 1911, Senator Jacques Catalogne proposed a bill that would reform the regulations governing the importation, commerce, possession, and use of opiates, and increase the fines to up to ten thousand francs and prison sentences to a year. In May 1913, members of the Chamber of Deputies followed suit, proposing three different bills that would have strictly forbidden the sale and circulation of opiates other than for medical usage. Radical socialist deputy Félix Chautemps, the sponsor of one of the bills, warned that "a large number of naval officers and colonial functionaries . . . are now smoking opium in France," and that the peril was great "not only for the individual victims of the awful habit, but also for the entire nation, which requires many of these people to serve the public good." Another radical deputy, Charles Leboucq, agreed, arguing that there was "no exaggerating" how important it was to check the flow of opium, particularly in the military. "The propagation of this vice could compromise the safety of the nation itself." The best solution, he reasoned, would be "to be pitiless towards those who forget their duty, for any gentleness towards them would be a crime against the *patrie*." Consequently, Leboucq proposed raising penalties to a fine of up to ten thousand francs and up to five years in prison for trafficking. These proposals did not make it out of the committee stage for several years, but they had widespread support in both the Chamber and the Senate.[76]

Increasing legislative anxiety over the problem spurred law enforcement to consider how it could make progress in its battle against the traffic in opiates. In the late spring of 1913, the Ministry of the Interior wrote to public prosecutors across the country asking for their evaluation of current opium controls and their suggestions on what could be done to make them more effective. In the prosecutors' responses, one common theme was that despite thorough investigations into opium dens and petty street dealers, the authorities were unable to figure out where contraband drugs were stored, leaving them powerless to identify, then arrest, the men and women who organized large-scale trafficking operations. Authorities' attempts to use the identification of local dealers and den managers to work their way up the supply chain and identify major dealers had failed, and according to some prosecutors, a new method for penetrating smuggling networks was essential. The major reason the bottom-up approach did not work, the Aix prosecutor explained, was that the penalties were not stiff enough to frighten petty dealers into handing over the names of their bosses. "Above all, [the laws] need to be strengthened," he argued, since "the fine that can be given now is not enough

to discourage those who take the risks" involved in opium dealing and trafficking. If the potential prison time were increased from two months to two
years, the prosecutors suggested, pushers would think twice before becoming
involved in the opium traffic. For recidivists, they suggested going even further, calling for provisions that would banish dealers from cities where they
had well-established smuggling networks that they could rejoin after serving
prison terms.[77]

The prosecutors also argued that no matter what law enforcement did in
its investigations, its efforts would be in vain as long as smugglers could get
drugs into the country with relative ease. The Rennes prosecutor complained
that customs agents "do not know very much about this drug; they are,
therefore, not well-informed about the importance of stopping smuggling,
or those who actually do the smuggling." To motivate customs officials, he
suggested monetary rewards for agents who uncovered traffickers trying to
sneak opium into the country. In a similar vein, the prosecutors argued that
postal officials needed to be more thorough in their searches of packages sent
from abroad, to choke the supply lines used by international smugglers.[78]

Most important, however, was the need to make the wording of the law
more explicit, since many judges had different ideas of what constituted an
offense against the 1908 decree. One judge in Paris, for example, found a man
caught smuggling opium in a train station innocent because the authorities
could not prove he had not purchased it legally. Other judges in the Aix and
Rennes jurisdictions refused to convict individuals found in illegal possession
of the drug, since possession was covered only indirectly by the 1908 decree.
By adding provisions specifically making it illegal to either push or possess the
drug, the prosecutors suggested, ambiguities in the law could be eliminated,
allowing more consistent and severe punishment for all offenders. Officials
in the Ministry of Justice echoed the concerns of the prosecutors, agreeing
that it was a stretch to use the 1908 decree to address the problem, given that
it focused on commerce and exchange and addressed possession and use only
indirectly. The only way to extend the reach of the law to cover users as well
as dealers, they said, was to create provisions specifically designed to regulate
not just commerce in the drug but also its use.[79]

For ideas on how to tackle the issue, government officials also turned
overseas, combing other countries' drug laws for ideas on ways to institute
controls that could be more effective than simple extensions of existing regulations. The nation with one of the most rigorous and draconian control regimes at the time was Japan, which proved particularly intriguing for French

administrators as they considered what measures could be taken. Forged over the course of the late nineteenth and early twentieth centuries by the Meji regime, Japanese drug control legislation featured some of the industrialized world's tightest controls over drugs, and some of the stiffest penalties for violations.[80]

Articles 136 through 140 of the Japanese penal code called for penalties of up to seven years in prison for individuals who imported, manufactured, sold, or possessed smoking opium or smoking equipment, and equally severe penalties for those who ran opium dens. The Japanese code also called for the same punishments for customs and law enforcement officials who turned a blind eye to smuggling and trafficking, a provision that French Justice officials imagined could effectively stamp out the corruption that sometimes hindered control efforts. Blending severe penalties that would dissuade both would-be smokers and dealers while unambiguously broadening the gamut of opium-related offenses, the Japanese control regime was, as Minister of Justice Aristide Briand summed up, "very interesting" as a model that the French could integrate into their own legislation. Officials in the Ministry of the Interior, though initially reluctant to transform the task of opium control into more of a police matter than it already was, soon echoed Justice officials' conclusion that the most promising model for the French to look to was that of the Japanese. "In reality . . . the most effective way . . . to strike at the managers of opium dens and those who regularly use these drugs is to aggravate the penalties," they wrote. "We cannot forget that this is a *penal matter* and that the law must be applied strictly."[81]

Upping the Stakes: The Law of 1916

The outbreak of World War I in the summer of 1914 provided the impetus for authorities to turn their tough talk about the drug problem into tough action. Opiates were not the sole targets of the legislative and administrative onslaught against intoxication that emerged during the war, as they became integrated into a broader campaign against alcoholic drunkenness. As soon as the conflict began, the government started cracking down on the use of France's most prominent psychoactive substance by instituting strict controls on the sale and consumption of hard alcohol, punishing public drunkenness, and banning absinthe. Given the importance of maximizing productivity and avoiding intoxication on the front lines, the war necessitated such measures to combat what historian Patricia E. Prestwich terms the French alcoholic "enemy within"—one that threatened to corrupt and incapacitate the war effort, which required the full focus and energies of the nation.[82]

Although opiates were integrated into the general wave of concerns about productivity and drunkenness that emerged during the war, their association with poor citizenship and treason gave them a special place within France's wartime anti-intoxicant crusade. The established links between opiates and Germany multiplied concerns about the drugs and fed suspicion that they were part of a Teutonic plot to compromise the French war effort. The French feared that alongside spies and other underhanded schemes, the Germans would use their pharmaceutical industry to wage a covert, psychoactive form of chemical warfare. "It seems that the Germans cannot defeat us with their fire or their asphyxiating gas," Deputy Charles Bernard speculated in February 1916, "so now they are using . . . morphine to try to wear us down."[83]

For some lawmakers, such as physician and senator Paul Cazeneuve, production of the drugs in German factories was proof positive that the proliferation of addiction in France was part of a broader German plan to destroy the nation. "Let's not forget that the factories on the other side of the Rhine send us . . . their morphine," he warned. Others in the Senate concurred, maintaining that it was, above all, "les Boches" who were feeding France's drug market as part of one of their most "dangerous invasions," a campaign described by socialist commentator Leo Poldès as an underhanded plan to "inundate the capital with Prussian poisons." Journalist Henry Rigal agreed, claiming that the scourge of opiates in powdered form was a distinctly German phenomenon: "The introduction of morphine into French medicine was a German one," he asserted, part of a broader "German plan" to achieve "the extermination of the French race by all means." Even doctors who specialized in treating addiction joined in the Germanophobic chorus, arguing that the importation of morphine into France must have been a German "prewar maneuver" designed to compromise France's military and patriotic prowess. Similar theories abounded in the mainstream press, some of which referred to addiction as a "new German scourge" that was "waging a much more dangerous war against us than the Kaiser's soldiers."[84]

The overlapping of concerns about drugs and the war endowed the opiate question, by 1915, with a new power and sense of urgency; the people who dealt in and used these drugs were no longer simply ordinary delinquents: they were threats to the nation. Consciously or inadvertently doing the enemy's dirty work by contributing to the destruction of national morale and devotion, those involved with drugs were not only sick and immoral but enemies of the state. Thus the French authorities needed to take all precautions necessary to ensure that the drug problem did not evolve into a crisis of truly national proportions.

To eliminate the danger that narcotics and those who spread them could pose to the national defense, the state took a logical step: it isolated them from the war effort. Under an 1849 law on governance during a siege, the government had the right to expel anyone considered a "dangerous individual" from regions where battles were being fought or the military was staging operations. Once the war began, officials targeted foreigners of German and Austrian descent, with forced transfer to detention centers in the western and southern parts of the country—something of a cross between displaced-persons facilities and prison camps. Non-nationals were not the only ones forced out of Paris and the areas near the fighting. Antimilitarists were often expelled from the war zone, as were prostitutes who were unruly or were rumored to be spreading venereal diseases. And as historian Jean-Claude Farcy found in his study of the evacuation and internment program, one of the most common reasons for Frenchmen to be expelled from the war zone was involvement in drug dealing.[85]

The archives of the Ministry of the Interior reveal that drug peddlers were specifically targeted "undesirable populations," especially in Paris. Beginning in October 1915, the authorities began keeping track of all drug arrests in Paris and, in most cases, called for offenders to be expelled from the capital. According to some writers, including Poldès, such decisive action against both users and dealers was a godsend, a "wonderful initiative" that promised to effectively keep narcotics away from the front lines. "Dangerous from a national point of view," as one police official described them, drug users and dealers posed a great danger to France and, as such, were to be treated in the same manner as pacifists and even Germans. Thus the place of opiates within the French nation reached a new nadir during World War I, as their image as agents of treachery—which had been established in colonial literature and in the Ullmo Affair—manifested itself concretely through the wartime evacuations.[86]

Despite these moves against the threat of opiates' poisoning of the war effort, intelligence reports of drug users remaining at or near the front underscored the need for more enthusiastic action. In 1915, for example, the Minister of War discovered that servicemen stationed near Paris were regularly attending clandestine opium dens in the capital, then taking drugs back to the front lines. This was worrying, not only because opiate-addled soldiers would be poor fighters, but also because introducing the drug into the barracks had the potential to spread the "contagion" of the opium habit throughout the armed forces. Once soldiers "bring their . . . habit to the front," the Minister of the Interior explained, "they will continue to procure opium, and lead their comrades to smoke it too." In the spring of 1916, revelations about a nurse who was working with the drug smuggler Lardenois, secretly supplying men on the battlefield, and sending military stores of opium to addicted soldiers stationed across the country, reinforced fears of the drug compromising the war effort. Despite the authorities' best efforts to use the 1908 decree to keep opiates away from the military, it seemed that the drugs, and some of their most prolific purveyors, had made inroads into the trenches. To protect the war effort from opiates and their demoralizing effects, then, more drastic action was needed. "It seems that the best way to really suppress this evil would be to destroy it at its root," the Minister of the Interior wrote to one of his subordinates in the summer of 1916. By "mercilessly going after the managers of opium dens as well as the merchants and dealers of the drug," he hoped the habit could effectively be weeded out of the armed forces once and for all.[87]

Around the same time, the Senate committee that was considering the propositions put forth in 1913 finished its deliberations and lent its support to the calls for a decisive legislative strike against opiate habits in France. "Parliament must resolutely face the reality, and with efficient and protective legislation, extirpate the evil," the Senate committee concluded in July 1915. The law that the committee proposed, however, added little to the existing control regime other than increased punishment—a maximum fine of ten thousand francs and up to two years in prison, as well as a retraction of civil liberties for five years, seizure and closure of premises used for opium dens, and doubling of penalties for repeat offenders. In November, the committee added cocaine and hashish products to the list of substances to be controlled, bringing its proposed legislation in line with the recommendations of the Hague Convention and addressing a growing concern about cocaine use in the military. When debate on the legislation began in early 1916, the response from members of both the Chamber of Deputies and the Senate was overwhelmingly positive, even earning enthusiastic applause when the proposal was presented. "The time has come to completely change our laws of the past," Deputy Bernard proclaimed in February 1916 in support of the proposal, urging his colleagues to institute a control regime and penalties appropriate for "our time" and for the "monsters" who were running the wartime underground drug market.[88]

Only one legislator, Senator Félix Goy, voiced any public opposition to the proposed law, expressing doubt as to whether such controls would be effective, anxiety about the law's wording, fear that the controls could make it more difficult for doctors to use opiates to relieve pain in their medical practice, and concern about unanticipated effects on the pharmaceutical industry. However, Goy expressed little concern for the plight of addicts in his oratory against the proposed control regime. If anything, it was a lack of concern for them that made Goy reluctant to support the law. Addicts "are all degenerates," he charged, "physically and morally stained." They were "not worthy of our concern," and to create a confusing and potentially detrimental set of restrictions just to protect them from their own narcotic appetites was, according to Goy, inadvisable.[89]

Nonetheless, a new poisonous substances law was soon passed. The government announced the law, appropriately enough, on Bastille Day, 1916, and put it into force with a decree in September. Many provisions of the new law simply strengthened existing regulations from the 1845 and 1908 control regimes, reiterating the requirement that individuals doing business in poisonous substances must register with the local authorities and making it illegal to sell drugs to customers without a prescription from a doctor, dentist, midwife, or veterinarian. For pharmacists, the poisons register requirements remained largely the same, and they were forbidden to fill prescriptions for opiates, cocaine, and hashish for more than one week's worth of drugs—a provision designed to keep users and potential dealers from stockpiling large amounts of controlled substances. In addition, pharmacies could now sell the drugs only to individuals who could prove they lived in the town where the business was located, unless the purchaser was from an adjacent village that had no pharmacist of its own. Exporters also became subject to tighter regulations, as they were required to obtain a customs certificate before shipping the drugs overseas.[90]

Perhaps most significantly, the new law broadened the gamut of prosecutable offenses. Article 31 of the September 1916 decree stipulated that possession of controlled drugs without a medical prescription constituted a violation, thus making it easier for law enforcement to make the case that individual users violated the law by "illicitly" possessing controlled drugs. As under the 1908 opium control regime, "encouragement" of use, whether by giving the drug away, by procuring a locale for use, or "by any other means," was subject to the same penalties in place for dealing, importing, and possessing drugs without a prescription. The law also made "usage in society" an offense, meaning that individual users with medical prescriptions who took the drugs in their own homes could escape the reach of the law, but those who consumed them in the company of others, particularly in opium dens, could not. Thus, many of the loopholes that had given judges leeway to let off defendants in cases involving use and possession under the 1908 decree were now eliminated; the new law clearly made these offenses punishable in most cases.[91]

The 1916 legislation clearly made the task of drug control a police matter more than a public health initiative. The September decree stipulated that local authorities and police, not medical men, were to oversee the surveillance and enforcement of the new regulations, and a new, dedicated drug squad was formed within law enforcement. The stiff penalties that violators of the new law would face—fines of up to ten thousand francs, prison sentences of up to two years, and a possible suspension of civil liberties for up to five years—reflected the shift to an increasingly law-and-order approach to drug control. The penalties for recidivists were doubled, and in cases of opium dens or other establishments where drugs were consumed or sold, authorities could confiscate the entire premises for up to two years (or four years for recidivists). As the public prosecutors had hoped, these provisions allowed the authorities to strike particularly hard at opium dens and other such establishments, not only by fining and imprisoning the individuals who ran them, but by taking their homes, businesses, and possessions.[92]

Supplementary regulations completed the control edifice set up by the 1916 legislation. In 1918, the military issued tighter controls over hospitals and pharmacies in the armed forces to prevent the diversion of opiates needed on the battlefield to the black market. In the early 1920s, when law enforcement complained that drug dealers were returning to their old haunts after serving their sentences and that it was too difficult to gather evidence in drug cases, the government passed a law allowing the ten-year expulsion of drug law offenders from cities where they violated drug laws and giving the police power to carry out surprise searches of suspected opium dens at all hours of the night. Later legislation further increased the punishments, with penalties for drug traffickers of imprisonment for up to five years and fines of up to thirty thousand francs. Thus, by the mid-1920s, the French government had instituted a new, more comprehensive, and harsher drug control regime to take the place of the 1908 decree and its relatively weak punishments. The Republic that had proclaimed itself the home of *liberté*, *égalité*, and *fraternité* sanctioned the isolation of individuals who violated drug control regulations from both their local communities (by banishing them) and the national community (by denying their civil liberties). In short, if a Frenchman broke the new drug laws, he could be stripped of the rights and privileges of citizenship in the Republic.[93]

According to Ministry of Justice statistics, in the early 1920s and 1930s, an average of 280 individuals were brought up on charges related to the 1916 law each year, peaking at 536 cases in 1930. In 1920 and 1921, a little more than half of the accused were either acquitted or let off for extenuating circumstances or for cooperating with police investigations. Those found guilty during this period were often given prison sentences, usually between three months and a year. Beginning in 1922, a higher proportion of individuals brought to court on drug charges were found guilty. Judges became more forgiving when meting out punishment, however; about half of the guilty verdicts for drug offenses in 1922 and 1923 resulted in more moderate penalties—either a fine or less than three months of incarceration.[94]

Historian Emmanuelle Retaillaud-Bajac's detailed study of drug cases in Paris between 1917 and 1937 helps shed more light on the nature of drug law enforcement in France during the interwar period—though, due to the unique characteristics of the capital, the story she tells may not be indicative of trends throughout the country. In Paris, the average punishment was more than four months in prison; 28.5 percent of those found guilty were fined between a hundred and a thousand francs, and 39.2 percent had to pay fines ranging from one to two thousand francs. More than two-thirds of those convicted for drug offenses in the capital during this period were male, and the majority of defendants worked in the medical field, the arts, or manual labor, or were unemployed. Laborers and artists were the most likely to be incarcerated, and for the longest time; doctors were more likely to pay heavy fines in lieu of prison time.[95]

Even though the heaviest punishments were generally reserved for traffickers and dealers, Retaillaud-Bajac's research reveals that those found guilty of offenses more likely to be linked simply to addiction—drug use and possession—still accounted for about half of the cases before the courts. Individuals involved in smuggling and peddling, however, were the most likely to go to prison on drug charges: more than ninety percent of them were incarcerated, for an average of more than six months. But the number of people imprisoned for offenses related to possession or use was not negligible: more than two-thirds of those with charges related to drug use served time in prison (for an average of two months), while those guilty of offenses related to possession had a more than seventy-five percent chance of going to prison (for an average of almost three months). The only punishment reserved for traffickers was banishment from their home cities: more than fourteen percent of traffickers faced this punishment, compared with less than three percent of individuals charged with possession, and less than one percent of those guilty of offenses related to drug use. After 1925, Retaillaud-Bajac finds, judges became more lenient with users than traffickers, tending to fine them instead of sentencing them to prison. Nonetheless, a large proportion of addicts did face harsh penalties for drug law violations and remained on the radar of law enforcement officials throughout the interwar period. Thus the legal provisions that allowed the arrest and prosecution of both individuals who dealt drugs and those who used them without a medical prescription were no hollow threat against the drug-using population of France. The law was used not just to limit the circulation of controlled substances on French territory but also to punish the individuals who used them. As Retaillaud-Bajac explains, though the 1916 control regime focused on repressing the drug traffic, it also "put a tight net around users, and multiplied the chances that they would encounter the police or judges, appear in court, or go to prison."[96]

Nonetheless, police remained frustrated with the punishments meted out by judges, many of whom rarely inflicted maximum sentences. And even after the cessation of hostilities in the fall of 1918, government officials remained worried that the Germans were still trying to poison France by inundating the nation with drugs and spreading addiction. As before the war, there was a grain of truth behind these suspicions. The economic troubles of Weimar Germany made a wide array of products—including narcotics manufactured by German pharmaceutical companies—particularly cheap, and smuggling out of Germany was a lucrative proposition for contrabandists of all stripes. Many of the drugs that found their way onto French streets were indeed manufactured in Germany, but the Franco-German border was a less common entry point for narcotics than were the Swiss, Italian, Spanish, and Belgian frontiers, or the French coastline. Few drugs were seized crossing the Rhine after 1919, and few German nationals were caught running major interwar smuggling operations in France. Yet both the medical establishment and the government remained convinced that Germany was somehow still orchestrating the French drug traffic.[97]

In December 1922, doctors Marcel Briand and Louis Livet of the Société de médecine légale de France claimed that drug smuggling into France was a legacy of the "demoralizing propaganda organized during the war" by the

Germans, who "continue, during a time of peace, to wage an underhanded war." Law enforcement officials echoed this sentiment, legitimizing the fear that narcotics were still being used as a Teutonic tool to destroy France. In the early 1920s, the Minister of the Interior made Germany the focus of France's drug interdiction efforts, tightening controls and increasing surveillance along the German frontier. In March 1921, he explained to his subordinates why it was particularly important to keep an eye on France's vanquished foes. Traffickers, he claimed, were now "being supported by the German authorities" who, by sneaking narcotics into the country, were working to "intoxicate" the French nation. To face the challenge posed by German would-be smugglers, customs officials were instructed to carry out particularly rigorous checks of all ships that arrived in France after docking in German ports, as well as thorough body and baggage searches of all passengers and chemical tests of all powders that were on board. Even though the conflict was over, French officials kept alive the wartime association between narcotics and France's main enemy, not just with their words, but with their actions. The jingoistic fervor of the war may have waned, but French nationalist discourses and practices surrounding narcotics remained.[98]

As the authorities focused much of their attention on the possibility that narcotics could be coming in from Germany, the drug traffic through France's other frontiers flourished. By 1922, narcotics became more than just another form of contraband smuggled by individuals trying to avoid customs duties. It was increasingly the province of crime syndicates involved in more nefarious activities. Expensive luxury items such as silks, highly controlled goods such as firearms, illegal products such as contraceptives, and even women trapped in the sex trade were all smuggled across France's borders, along with narcotics, meaning that the groups behind the underground drug market were no longer amateur dealers or smugglers. Clearly, these were not petty smugglers but organizations involved in a wide range of dangerous and illegal enterprises. In spite of the stiff penalties called for by the 1916 law—or perhaps because of them—the drug traffic became increasingly lucrative, a trade whose potential profits outweighed the possible penalties for a good number of criminal groups operating in France.[99]

Many of the narcotics smuggled into France, however, soon made their way out again. In large part, this was because the French had not yet adopted (and did not do so until 1928) the system of import-export certificates recommended by the League of Nations, making it relatively easy to smuggle narcotics out of French ports. What is more, unlike the British, the French did not place limits on which ports could export opium, or to which countries, meaning that it was more difficult for the authorities to track who shipped the drug where or to ensure that narcotics did not flow from French territory to the international black market. Furthermore, smuggling through France was probably exacerbated by the French authorities' gross inefficiency in detecting smuggled goods. The government estimated less than five percent of the narcotics smuggled through France were detected by customs. In addition, government regulators tended to be lax about tracking how much morphine, heroin, and cocaine French pharmaceutical firms manufactured, so quantities of French-made narcotics may have found their way to the black market. North America, in particular, was a common destination for the narcotics that passed through France. In the early 1920s, for example, a group of British, French, American, Canadian, German, and Swiss traffickers organized a global smuggling route through Marseille, using the port to ship narcotics manufactured throughout western Europe to the streets of the United States and Canada (as well as China and Britain). Marseille, in fact, became a global hotspot for many illicit traffickers, due to its regular contact with boats from Indochina and the Middle East. By the early 1930s, smuggling operations that were precursors of the "French Connection" of the postwar period were running out of Marseille, as a group of Corsicans began using the port as a transshipment point to move large amounts of drugs from Turkey (where smoking opium and heroin were manufactured) to North America.[100]

Although the new regulations were extremely ineffective at stemming the international traffic, many in France believed that they were achieving their principal goals: to stifle domestic drug dealing and transform France into a drug-free nation. Several studies in the late 1920s and early 1930s found that the scarcity of addictive drugs was an important factor in convincing many Frenchmen to quit using narcotics. In their 1922 speech before the Société de médecine légale de France, Briand and Livet directly credited the operation of the 1916 law—and how difficult it made procurement of narcotics—with reducing rates of addiction. French legislation, they argued, was among the "most severe" and "complex" in Europe, and a habitual drug user could no longer function in France. Whereas other countries such as Spain, Italy, Germany, Britain, and Switzerland were relative "lands of plenty" for addicts, the two doctors claimed, the scarcity of drugs in France ensured that Frenchmen would stay away from them: "This explains why and how our addicts are increasingly going to other countries where legal complacency makes it easier for them, and allow[s] them to freely engage in their harmful habits." Thus, by pushing addicts to either quit using drugs or leave the country to procure them, Briand and Livet concluded, the 1916 law was beginning to work, helping eliminate drug use from France.[101]

As addicts either renounced their habits or left France out of necessity, the logic went, the nation could eliminate the corrosive influence of narcotics and their users on its political and social bodies. Furthermore, with the array of penalties and sanctions against offenders under the laws of the 1910s and 1920s, the state had the weapons it needed to strip drug users of their liberty, their property, their civil rights, and even their ability to return to their home cities—in short, to deny them their citizenship. Thus, unlike in Britain, where concerns about economic well-being colored the development of drug control initiatives and policies, in France it was the goal of creating a citizenry that did not overindulge in narcotics that underlay the drug policies of the early twentieth century.

The Motors of Drug Control: Economic Exigency and National Security

The major moves toward opiate control during the World War I era were initiated to address a particular set of concerns related to the economy and the war effort in Britain, but the logic undergirding the repressive opiate control measures in France was more enduring. For the British, the need to suppress smuggling that endangered commerce became less immediate with the cessation of hostilities, attenuating the necessity of strictly enforcing drug regulations at home. This, combined with an apparently diminishing number of illicit drug users, led the number of arrests for violations of the Dangerous Drugs Act to become exceedingly small by the mid-1920s. In France, by contrast, even after the immediate crises of war and national security became less pressing following the Armistice, the campaign against narcotics remained an ideologically charged and nationalist one throughout the interwar period. Given that French drug discourses, dating back to the late nineteenth century, were more closely intertwined with concerns that opium use was irreconcilable with tenets of solidaristic ideology and national identity, the survival of anti-narcotic nationalism well into the postwar period is hardly surprising.

Opiates, for the French, were not problematic for simply practical reasons of health or moral reasons concerning sobriety. On a philosophical level, opiates were the chemical quintessence of poor citizenship and anti-patriotism that posed a serious threat to the health and survival of the *patrie*. The task of creating an opiate-free nation, therefore, was more central and enduring in the political imagination in France than in Britain. The French attitude concerning opiates, and opiate users, was thus more hostile and intense than that of the British, and the discourses, policies, and penal arsenal developed in France's anti-drug campaign reflected this. In Britain, opiate control was largely a means to an end, designed with specific aims and policy goals that were largely attained with the end of World War I and the normalization of international trade. In France, on the other hand, drug control was an end in itself. The struggle was one of enduring philosophical importance for the French, as borne out by the persistence of politicized and Germanophobic thinking that marked drug discourses and policies before, during, and after World War I. A task with relatively discrete policy aims and goals in Britain, opiate control was a more enduring enterprise in France, driven not just by pragmatism but by political principle.

Each nation's anti-narcotic nationalism, elaborated in the nineteenth century, shaped the development of drug control initiatives in the twentieth, and would also influence how each country treated its addicted citizens as their narcotics control regimes took hold.

CHAPTER FOUR

Control and Its Discontents

The Plight of Addicts under Opiate Control

It is a pity that medicine does not try to make opium inoffensive instead of perfecting the process of detoxification.

—*Jean Cocteau*

Curing opiate addiction in the late nineteenth and early twentieth centuries, as physicians learned from their experience treating addicted patients, was not an easy proposition; it required a huge investment of time and resources and tremendous amounts of patience. After some fifty years trying to perfect a process for weaning addicts off opiates, most experts agreed that the chances of achieving a lasting cure were still slim, even in the best of circumstances. By the 1920s, when tighter drug control regimes were put in place in both Britain and France, this would create a conundrum that was not only medical but legal. Given that addiction was often incurable, but drugs were exceedingly difficult to obtain legally, what was an addict in the age of tighter narcotics control to do? British and French responses to this question were radically different. Differences in the perceived scope of their drug problems and in their conceptions of addicts' place in the national community would play key roles in determining how Britain and France treated their addicted citizens.

Curing Addiction: An Impossible Task?

As they became familiar with opiate habits, medical and psychiatric researchers realized that one of their defining characteristics was that they were extremely tough to break. Confronted with the challenge of trying to cure addiction, many late-nineteenth-century doctors and psychiatrists devoted a large portion of their research careers to figuring out the best treatment approaches, but to little avail. The next generation of practitioners enjoyed equally little success, and by the 1920s, the art of weaning addicts off opiates had hardly advanced beyond its position in the 1870s. The best way to treat addiction remained, as Norman Kerr summed up in 1895, "a subject of considerable controversy." More often than not, the British and French therapeutic communities were left to combat the seemingly invincible foe of opiate addiction with perseverance, luck, and not much else.[1]

The Tools and Tricks of Addiction Treatment

Even when opiate addicts entered treatment willingly, they began to resist—sometimes violently—as soon as they started suffering the symptoms of abstinence. Anxiety and withdrawal pains threw patients into a tenuous state, alternating between extremes of physical distress, delirium, depression, and suicidality. To make the process less traumatic, practitioners tried easing patients off opiates by habituating them to a new substance, such as cocaine, alcohol, cannabis, or even an opiate other than the one the patient was addicted to. The logic here was that these other drugs could serve as "transitional medicines" and soothe the cerebral and nervous cells that had been habitually stimulated by opiates. Around the turn of the century, however, most doctors began to avoid this practice, finding that alcohol and other drugs tended to aggravate, rather than attenuate, the symptoms associated with withdrawal. Furthermore, even when helpful in the short term, such substitution practices made the cure a longer, more drawn-out process, as the dose of the new poison would have to be tapered off once the patient had become opiate-free. At times, depending on the substance used and how long it was administered, doctors found that patients would develop full-blown addictions to the substitutional drugs, thus creating new—and sometimes much more serious—problems. The clearest examples of this came in the late nineteenth century, when practitioners began using two new drugs to help wean addicted patients off morphine. The first was cocaine, praised by many practitioners—most notably Sigmund Freud—as a potential cure for morphine addiction in the 1880s. The second was heroin, an opium derivative first marketed in the late 1890s. Soon enough, doctors began to recognize that both of these substances could exacerbate addictive behavior rather than cure it and that their habitual use could, in itself, soon develop into a disorder.[2]

Given the danger of escalating addiction with substitution treatments, most doctors turned instead to the remedies they used in the treatment of mental illness to make the withdrawal process go as smoothly as possible. For the most part, this meant giving chemical and physical treatments designed to sedate while easing discomfort. Oral, rectal, and subcutaneous administrations of emetics, ether, alcohol, water, milk, coffee, tea, soup, calomel, ammonia, insulin, chloroform, chloral, bromides, hyoscine, and belladonna were used in combination with warm or cold showers and baths, massages, electricity, and ultraviolet light to cleanse the body of opiates, provide pain relief, and keep patients under control. Many practitioners created their own mix of dietary and chemical regimens, but none proved effective or reliable enough to become a standard. Some adopted more unorthodox methods to help addicts through withdrawal. For example, Oscar Jennings, a specialist who treated addicts in both Britain and France, used everything from gramophones to prolonged bicycle rides to keep addicts' minds off their withdrawal symptoms, with the hope that distraction could get them through the most painful parts of the process. Yet no matter what method was used, addicts were prone to becoming agitated or violent when denied their drugs. The temptation to go back to opiates was great, and unless compelled to stay the course, many addicts gave up on the treatment before it could be completed.[3]

Most practitioners maintained that, for treatment to be effective, it needed to take place in a highly controlled environment, one where the addict could receive immediate care and be closely monitored. Some of the early experts on opiate addiction, such as Edward Levinstein, laid out specific guidelines for how the addict should be treated and controlled once institutionalized. For the first three days of withdrawal, Levinstein recommended having a medical attendant on twenty-four-hour duty in an adjoining room, armed with a reserve of interventions—chloroform, ether, ammonia, mustard, ice, and an electric induction apparatus—that could be used to calm, sedate, or resuscitate the patient in the event of an emergency. Severe symptoms were the norm, Levinstein warned, so caregivers needed to be prepared for addicts to collapse, enter shock, or have seizures, and ready to respond by using "counter-irritation" techniques such as administering smelling salts, shaking patients, yelling at them, and even lashing them to keep them awake until an emergency dose of drugs could be provided. Sometimes these methods fell short, and on occasion, patients died from complications related to withdrawal—even while under the close care of physicians trying to cure them.[4]

The occasional emergency resuscitation was just one of the unpleasant responsibilities of those charged with caring for addicts in treatment. Medical attendants also had to be judicious in their interventions, not dispensing drugs unless they were absolutely sure the patient was truly in dire need of them. According to many medical researchers, judging this was often difficult, because addicts were willing to fake medical crises to get a dose of their drug of choice. Caretakers, they warned, needed to be prepared to handle such ruses by refusing even the most desperate and heartbreaking pleas of the patients under their watch. Thus nurses and orderlies faced a doubly difficult task: they not only had to be ready to perform dramatic interventions to preserve the well-being of their patients; they also had to maintain a healthy suspicion of them. A mistake in either direction—giving drugs when they were not necessary, or refusing them when they were—could jeopardize the withdrawal process or cause irreparable harm to the patient's health. Treating the addict, therefore, was no ordinary task. It required a high level of training, judgment, and discernment, and it was work that could be taxing both physically and emotionally.[5]

Beyond trying to deceive their caretakers into giving them drugs, researchers maintained, addicts were also prone to subverting their doctors' efforts and sneaking opiates into treatment with them. Practitioners reported that patients would hide vials of morphine solution underneath their mattresses, in the yards of treatment facilities, or anywhere they could hollow out a secret niche—in a couch, a bed, a bureau, or whatever else was available. To prevent patients from sabotaging their own recovery, therefore, caretakers would need to don the cap of the prison guard, regularly checking the furniture, belongings, and sometimes even the body cavities of those they treated. Some experts recommended making it clear as soon as an addict entered treatment that, for his own good, he would have to surrender his freedom and, as Levinstein said, "submit, without opposition, to the orders of the medical attendant." To this end, they recommended making entrance into treatment similar to matriculation into a prison. Georges Pichon, for example, made it a policy to search patients when they entered treatment and to bathe them and give them special clothes without pockets or other areas where drugs could be hidden. To prevent the smuggling of drugs into treatment, Pichon also forbade visits from the outside and instructed guards to thoroughly search all packages that patients received by mail. Other practitioners recommended going even further, putting patients in specially designed rooms with nowhere to hide drugs and nothing that could be used—should the suffering of withdrawal become too great—to escape the facility or commit suicide. Windows were to be more than an inch thick so that they could not be broken; doors and windows were to move on pivots instead of hinges so that they could not be disassembled; doors were to be constructed so that patients could not open or shut them; walls were to be padded; and hooks and curtains were to be removed so that they could not be used to inflict self-harm. Most importantly, practitioners emphasized, furniture needed to be specially designed so that it could not be taken apart to create tools, weapons, or hiding places for stashes of drugs.[6]

Although not all practitioners took such drastic measures, the logic behind most addiction treatment practices of the day did focus on controlling and containing the patient's impulses while he underwent withdrawal. Whether through chemical sedation, mental distraction, or physical restraint, all of these strategies were designed to keep patients from acting out and to get them through what most agreed was the most arduous stage of treatment—detoxification.

The Art of Weaning

The main issue that divided the therapeutic community was not so much what treatments should be given but how fast the weaning process should proceed. Should it take days, weeks, or months? The question was critical; doctors and psychiatrists maintained that taking addicts off drugs too quickly or too slowly could determine whether treatment would lead to a cure or a relapse. Levinstein was among the strongest advocates of immediate withdrawal, and some practitioners followed his lead. Though brutally painful, forcing patients to abstain immediately was best, Levinstein argued, because it maximized the likelihood of a lasting cure. If a patient's dosage was cut back in a gradual or iterative process, withdrawal symptoms might be less dramatic, but they would also last longer. An intense seventy-two-hour period of discomfort was a fair price to pay, he argued, if it could prevent more prolonged suffering in the long run. A more "gradual deprivation," on the other hand, would ultimately lead to greater aggravation, since patients would suffer anew every time their dosage was decreased. The mental anxiety caused by anticipation of decreasing doses was likely to make patients "fretful and irritable" to the point that they might get discouraged and quit treatment altogether. Thus, to use the gradual withdrawal method, he concluded, was to make a deal with the devil, exchanging the brief, though acute, suffering of immediate withdrawal for a more protracted ordeal that was, in the end, more physically and emotionally taxing.[7]

Levinstein also argued that if the withdrawal process was long and drawn out, patients would come to distrust their doctors. For the duration of time that the addict was being weaned off drugs, the relationship between patient and doctor was an adversarial one, with the caretaker holding the power to ease suffering by providing opiates, but refusing to do so for the patient's own good. The longer the withdrawal process took, the longer the doctor would have to play the role of disciplinarian by denying the addict the drugs he craved. Consequently, the doctor-patient relationship would erode, and the patient would gradually become less likely to follow the doctor's instructions. Thus, in Levinstein's view, for reasons that were both medical and psychological, sudden withdrawal was preferable because it was the best way to minimize the addict's mental anguish and foster a healthy, trustful relationship between patients and their caretakers.[8]

Though decried by many as cruel and callous to addicts in withdrawal, Levinstein's line of thinking had some adherents in Britain—and more in France—throughout the late nineteenth and early twentieth centuries. Experience with addicts who had been almost completely weaned by a more gradual method, but then were unable to take the last step of becoming drug-free, frustrated many doctors and led them to believe that the swift and decisive, albeit painful, process of immediate and complete withdrawal was the only reliable one. Those who criticized sudden withdrawal methods, wrote Daniel Jouet of the Paris Faculty of Medicine in 1883, had not understood that more gradual approaches would unnecessarily "drag out" the difficult process of curing addiction. Many also agreed with Levinstein that, despite their merits, more gradual withdrawal methods gave what Aleister Crowley termed a "horribly disquieting impression that the patient could not trust the doctor," and that their use could do irreparable damage to the doctor-patient relationship.[9]

Practitioners were not alone in their critiques of prolonged withdrawal methods, as became clear in Crowley's semi-autobiographical *Diary of a Drug Fiend*, published in 1922. More gradual withdrawal, he claimed, tended to breed mistrust and avarice in the addict, who would come to see his caretakers as "ignorant and heartless charlatans" who made the process torturously slow. To undergo a treatment in which "the daily amount was reduced by a series of jerks," Crowley claimed, was akin to being "sentenced to be flogged at irregular intervals without knowing exactly when." This left the addict "in a state of agonizing apprehension" that was more painful than sudden and decisive abstention from opiates. A French addict who experienced a gradual cure in the early 1920s agreed that the experience of gradual withdrawal was both horrific and ineffective, telling her doctor that "I had such memories of it that I would have preferred to stay a morphine addict."[10]

Such arguments and testimonies aside, by the beginning of the twentieth century, most in the British and French medical communities came to see the sudden withdrawal method as too dangerous and painful to be effective, and preferred more gradual weaning processes. According to critics of sudden withdrawal, the practice was irresponsible and hardly worth the risk, since it was so excruciating that it was likely to discourage addicts and push them to quit treatment. Furthermore, the potential complications from sudden withdrawal were much more severe than those associated with more gradual methods, and more likely to cause serious physical harm. Thus sudden withdrawal was, as French doctor Paul Sollier asserted, "pointlessly barbaric." F. S. D. Hogg, a British practitioner who specialized in treating addicts in the 1920s, was even more blunt in his assessment, arguing that it was "barbarous and inhumane to expose a person unnecessarily to dangerous severe shock and acute suffering." Any doctors willing to withdraw drugs from their patients in one fell swoop, he charged, did so simply "to save themselves trouble or because they are quite indifferent to the feelings of their patients."[11]

The critics of immediate withdrawal also thought that the suffering associated with the weaning process needed to be minimized for practical as well as moral reasons. If not addressed, the indescribable pain experienced during immediate withdrawal could be psychologically damaging and emotionally scarring, leaving the addict nostalgic for drugs and more likely to go back to them later. Addicts would become "so enraged" by the pains of immediate withdrawal, Maurice Page, director of a hospital near Paris, explained, that "they [would] swear the first thing they would do when they left [treatment] would be to get an injection." If a cure was too painful, Roger Dupouy elaborated, addicts would run back to drugs, not out of necessity, but to "avenge their suffering." Thus a withdrawal process that was as painless as possible was preferred because it would lessen the likelihood of relapse. Advocates of gradual withdrawal also argued that minimizing the pain would enhance the doctor-patient relationship, not weaken it as Levinstein suggested. "When the weaning has been affected . . . by compulsion instead of guidance," Jennings counseled, a "latent hostility" against the doctor would emerge and jeopardize the patient's ability to work with doctors in the future. Gentler and less painful methods would be more effective because, with a less unpleasant experience, patients who relapsed would be more willing to return to treatment.[12]

To avoid the tortures of immediate withdrawal, most practitioners turned to a more gradual withdrawal process that ranged from a few days to several months. There was no consensus on the specifics of how the tapering should be done. Some doctors recommended cutting a patient's dosage in half when he first entered treatment, then diminishing it by a set percentage until it neared zero, at which point a final withdrawal would be carried out over the course of a week. Other doctors maintained that the withdrawal process could not begin until the patient's health was stabilized under a doctor's care, so it could take even longer. Many chose not to have any hard-and-fast rules dictating speed, and they set individual timetables depending on the amount of drugs habitually consumed, the duration of the addiction, and the patient's health and character.[13]

Though many practitioners believed they were being more humane and respectful of their patients than were their colleagues who used sudden withdrawal methods, it was nonetheless a standard practice for doctors to deceive their patients during more gradual tapering processes. Often this was done by decreasing the drug concentration in each dose rather than the size of the dose, so that patients saw the same amount of solution being injected each time but did not realize that each dose was becoming progressively weaker. To ensure that the ruse was not compromised, some practitioners recommended lying to patients, telling them they were receiving more of the drug than they were actually getting. With this method, physicians could ensure that patients would "not [be] made aware of the actual moment when the drug has been totally withdrawn" and decrease the likelihood that they would protest or experience psychosomatic discomfort. To be sure that the patient remained oblivious to his progress, the British Ministry of Health recommended continuing hypodermic administrations of "innocuous solutions" for up to two weeks after the final withdrawal. Such benevolent deception, if done properly and with a sufficiently gradual tapering schedule, would keep the patient completely unaware that he was being weaned, and thus minimize resistance. "The patient is then surprised to learn that he is no longer taking the drug," the Ministry explained, and on realizing he was no longer receiving any opiates, he would consent to stop injections and be successfully weaned off his drug of addiction.[14]

After Withdrawal

Regardless of the method used for getting addicts to become drug-free, most practitioners saw the weaning process as just the first, not the ultimate, step of the therapeutic process. After detoxification, it was important to help restore the patient's health, since the physiological and psychological ravages of prolonged opiate use remained long after the addict stopped using drugs. In the 1870s, the focus of rehabilitation was largely physical, as practitioners advocated a strict regimen of long walks and other physical activity to create fatigue and restore normal sleep patterns without recourse to chemicals. As they gained more experience dealing with aftercare, practitioners began to adopt a more holistic approach to the restoration of physical and mental equilibrium that focused on more than just the attenuation of withdrawal symptoms. To fully rehabilitate the addict, experts such as Kerr recommended a convalescence period of several months filled with outdoor exercise and indoor physical labor, to be complemented by "elevating influences of music and the fine arts, intellectual culture," and "moral and religious considerations" that could nurse the mind and spirit back to health. Other practitioners agreed that an "after education" program of some sort was key, since it would strengthen addicts' willpower and decrease the likelihood of relapse. With complete and well-rounded "teaching and counsel," Kerr summed up, the practitioner could "inspire the former listless and despairing narcotic slave with new and noble resolves," while arming him with healthier options than the opium pipe or syringe when he was tempted to return to his old ways.[15]

Despite their best efforts, practitioners often failed to effect a lasting cure for their drug-addicted patients, even in the best of circumstances. "It is still difficult," French doctor Paul Constantin lamented in 1891, "even when the patient wants to be cured and makes a sincere effort." Some thirty years later, the prognosis had barely improved, as doctors regularly saw patients who had tried to quit up to three, four, or five times but were still unable to abstain from opiate use. In its study of addiction treatment in 1926, the British Ministry of Health found that failures in treatment were almost to be expected, because "relapse, sooner or later, appears to be the rule, and permanent cure the exception." Even in institutions that specialized in treating addiction, the chances of success were slim. Doctors at the Dalrymple House, one of Britain's leading addiction treatment centers, estimated that only fifteen to twenty percent of their patients would "do well" after discharge. In France, doctors who specialized in treating addicts at the Hôpital Henri-Rouselle reported slightly better numbers, giving a generous estimate that one in three patients who completed their program stayed clean and sober.[16]

No matter what method was used, addicts still found the withdrawal process perilous and torturous. Deciding between sudden and gradual withdrawal, Crowley mused, was akin to having a "choice between plunging into boiling oil and being splashed with it every day for an indefinite period." Given the low rates of success, the physician treating addiction seemed, from the perspective of the addict, little more than a "witch doctor with a license from government to torture and kill at extravagant prices." In his *Diary of an Addict*, published in 1930, Jean Cocteau was equally skeptical of the methods used to cure his opium habit at the neurological clinic at Saint-Cloud. His treatment regimen was, he mused, as precise and effective as "the legend that said that demons could be driven out with plants, charms, purges, and emetics." In short, he concluded, "scientific disintoxication does not yet exist."[17]

With the difficulties inherent in curing opiate addiction, addicted individuals found themselves in a precarious situation, to say the least, after the institution of tighter opiate controls in Britain and France. Compelled by their bodies to continue using drugs, yet cajoled by the law to abstain, what were addicts in the age of narcotics control to do?

The Plight of Opiate Addicts in Britain

In Britain, the consequences of drug control became abundantly clear to medical practitioners, addicts, and law enforcement officials by the early 1920s. Doctors were unhappy that the Home Office wanted to impinge on their right to prescribe opiates in their medical practice as they saw fit. Addicts were unhappy that they were being denied a medicine essential for their everyday functioning. And officials in the Home Office grew frustrated because many doctors seemed to be abusing their prescription privileges without being brought to justice. After extensive research in the mid-1920s, however, many of the shortcomings of the early control regime were rectified, and by the end of the decade, Britain had developed a new approach to the drug problem that answered, at least partly, many of the early critiques of its drug policies.

Options for Addicts

Because most addiction experts maintained that the best prospects for curing addiction lay in institutional care, many British opiate addicts found their way into inpatient environments in the late nineteenth and early twentieth centuries. Often this meant turning to the treatment options available to other individuals with substance abuse problems—namely, alcoholics.

Extreme cases of alcoholism were often treated under the broad gamut of care for the mentally ill and in Poor Law workhouses, but Britain began developing a distinct system for handling alcoholism in the late nineteenth century. Under the Habitual Drunkards Act of 1879, local authorities could grant licenses to "inebriate retreats," and under the Inebriates Act of 1898, the government inspected them yearly. Confinement in a retreat was, in theory, completely voluntary, though most alcoholics who entered these institutions did so under pressure from family and friends. The retreats were not funded by the state, so unless the local community or charities were willing to finance them, their services were available only to those who could afford them. The 1898 act, however, also authorized the creation of state inebriate reformatories, institutions partially financed by the central government and designed to serve as prisons to rehabilitate alcoholics. For individuals deemed to have committed crimes because of excessive drinking, courts could order their detention in these reformatories—rather than in prison—for up to three years. Once interned, alcoholics were compelled to serve out their sentences unless considered "cured," and if they escaped, the authorities had the right to arrest them and bring them back.[18]

Although some opiate addicts found their way into the publicly funded reformatories, a series of cases in the 1890s established that the institutions were to treat only those individuals who drank their poisons, meaning that many addicts—those who smoked or injected their drugs—fell outside their target population. Despite the wishes of those who specialized in treating addiction, injection-drug users were not covered by the Inebriates Act. The private homes that took non-lawbreaking drinkers, however, did accept many opiate addicts. Most notable among these was the Dalrymple House in Rickmansworth, which soon became one of Britain's leading centers for addiction treatment research. According to Dalrymple's statistics, about one in every thirteen individuals who entered the facility between 1900 and 1925 were opiate addicts. Since the private homes charged fees, however, their clientele was generally limited to the well-to-do.[19]

Addiction specialists sought to extend the provisions allowing the publicly financed treatment of alcoholics to cover opiate addicts, and they convinced the government to take steps in that direction in the first decade of the twentieth century. In 1908, a Departmental Committee on the Inebriates Act accepted that drug taking should be covered by the same set of rules as excessive drinking, though this concession came too late to have any practical effect on the treatment of opiate addiction. By the time the need for better treatment options became more pressing with restrictions on supplies resulting from provision 40B of the Defence of the Realm Act (DORA) and the Dangerous Drugs Act, the state institutions for treating alcoholics had basically become defunct. Local authorities were reluctant to foot the bills for the reformatories, which soon became overcrowded with patients who required more care than the institutions could provide. Consequently, by the early 1920s, alcoholics, as well as drug addicts, were once again cast into the broader treatment system for the mentally ill and the indigent. The wealthy could continue receiving services for their addictions in private institutions, but the state was unable to provide help for addicts until their condition became so serious that they could be classified as mentally ill under the 1890 Lunacy Act or as "defective" under the 1913 Mental Deficiency Act. By the early 1920s, British opiate addicts who did not have the means to finance their own treatment were left between a rock and a hard place: they had to find provisions to feed their habits illegally, find a sympathetic doctor to prescribe drugs for them, try to stop using on their own, or deteriorate so drastically that they would become a ward of the state.[20]

On occasions when addicts did end up in state institutions, the treatment they received was often inadequate. One addict put in a psychiatric institution in 1923, for example, became so agitated when she went into withdrawal that the staff decided to release her so she would not disturb the other patients. The only public institutions that reported any success in treating addiction, in fact, were prisons, where addicts were usually subject to immediate or very rapid withdrawal. Medical officers sometimes gave addicted inmates sedatives to keep them calm and well-behaved for their appearances in court; after that, inmates were given either no drugs or very small doses. According to one prison commissioner, this method was very effective, largely because, since prisoners were incarcerated, they had no way of procuring drugs and chose to reluctantly "assist in the treatment" rather than resisting it.[21]

Even though prison officials claimed that it worked, forced detoxification behind bars was hardly an optimal solution. Magistrates who saw addicted drug law offenders in their courtrooms recognized this, and many gave deferred sentences on the condition that addicts would promise to enter treatment, either as paying patients or with the assistance of charitable institutions. Government officials, though cognizant of the practice, were loath to use the courts to coerce addicts into treatment by threatening to imprison them. Many officials, particularly in the Ministry of Health, lamented the lack of options for the addict of modest means who had not run afoul of the law or become mentally ill. "No general hospital . . . takes or would take cases of drug addiction," Dr. E. W. Adams, of the Ministry of Health, wrote in the early 1920s. "There are a few private institutions which deal with the treatment of the narcotic diseases," but they were "only available for people with ample or at least moderate means." In 1924, another government official agreed, writing that the state could not make it a policy to force addicts into treatment, since most simply "could not afford the cost."[22]

Early Critiques of the Drug Control Regime

The apparent abundance of police measures but lack of health measures to address the problem of drug addiction attracted attention outside the halls of government by the early 1920s. W. E. Dixon, a leading addiction specialist, maintained that the current control regime was fundamentally flawed because, in practice, it transformed the medical issue of addiction into a legal one. "Cannot our legislators understand that our only hope of stamping out the drug addict is through the doctors?" he pleaded in an open letter to the *Times* of London in 1923. Since drug dependence was a health problem above all else, he maintained, "we can no more stamp out addiction by prohibition than we can stamp out insanity." The editors of the *British Medical Journal* agreed, suggesting in 1921 that "the cure for intemperance in narcotics . . . must lie in the development of character" rather than in the proclamation of laws, since "a way will always be found to get round coercive legislation." To use police as the main weapon in the fight against addiction, they argued, was not only ineffective but a violation of the Briton's right to freedom from state interference in the private realm. "To control conduct so that the individual has no liberty of choice," the editors warned, was "to court disaster."[23]

In the *English Review*, Crowley, writing under the pseudonym "A New York Specialist," echoed these sentiments. "When the course of nature is violently diverted by pious Puritans and profiteering policemen," the issues caused by addiction were bound to multiply. Were "heroin . . . as easy and cheap to buy as butter, the whole underground traffic would disappear like a bad dream," and the authorities would be able to gain control over the drug problem. The focus on cutting off supplies only spread rather than controlled addiction, according to Crowley, since it created incentive for criminals to join the underground market and work to recruit new addicts so they could gain new customers. Given that those who wanted drugs would "manage to get them one way or another," he concluded, all prohibition accomplished was to make it more difficult for drug users to do something they were going to do anyway. Thus a policy built around control, Crowley concluded, contributed to the "degradation of citizenship" by trying to use force to address a set of problems that would be better tackled with medical attention.[24]

Aside from doctors and intellectuals, ordinary citizens who were addicted to opiates also made their voices heard by writing letters to the Home Office complaining about the new drug control regime. One such addict was Thomas Henderson, a fifty-eight-year-old painter of miniature figurines and a photography teacher, who came to the attention of the authorities when they noticed he had been procuring large amounts of morphine from an Oxford chemist in 1922. Henderson had begun taking morphine daily at the age of twenty-one to treat a medical condition, and he tried to quit three times over the subsequent decades. But each time he stopped taking morphine, he found he could no longer work or support himself. As he wrote to the Home Office in November 1922, he needed the drug to function and make a living, and he did not use it for its psychoactive effects. For a man of modest means, with few savings and a family to support, taking the drug to stave off withdrawal symptoms was the only way he could remain self-sufficient. After passage of DORA 40B and the Dangerous Drugs Act, Henderson made a fourth attempt to stop using morphine but was unable to do so, as he became incapacitated by cramps and severe mental distress. When he reported this to his doctor, Henderson was specifically advised *not* to quit, since abstaining could potentially kill him or, at best, leave him a shell of the man he once was. Thus Henderson, an otherwise law-abiding and productive citizen, faced an unenviable dilemma: either continue taking morphine and break the law or go without, at great risk to his health and the well-being of his family.[25]

In his letter to the Home Office, Henderson did more than just make a desperate plea for pity: he also mobilized the language of British citizenship, particularly that of the economically productive and self-sufficient citizen who should be left to his own drug-taking devices. If denied morphine, Henderson claimed, "I am to be a wreck and unable to earn my living and pay my rates and taxes," whereas with it, he could be "perfectly fit in health and working power." It was *because* of morphine, not *despite* it, that he had become not just a passable citizen but a model one. With the drug, he maintained, "I have gone through a long life honestly and ably, never owing any man a penny, paying my tradesmen weekly as a rule, and never being in any man's debt." Having spent almost forty years building a business, giving photography lessons, and writing a book with the morphine syringe by his side, Henderson saw little reason that he should be compelled to give it up. "I have proved over and over again that I can give up the habit as a habit," he explained. "But it leaves me powerless to carry on my life's work, and so I am useless." Morphine and productivity were not enemies, therefore, but allies for Henderson. Given his history of self-sufficiency and productivity while on the drug, the government should have, if anything, wanted him to continue using it so that he could continue to be a productive citizen. "It cannot be the wish or the desire of the State to crush out the existence of an honest man who has originated a very original line of business," he reasoned. He could continue to lead a "useful life to the state" by teaching others to start successful careers, while also building his own, but only if he could have the drug he needed to function. His request to be allowed to continue taking morphine, therefore, was not simply an appeal to charity but, clearly enough, an appeal to the right of the British subject to live an orderly and productive life. All he wanted, he concluded in his letter, was "the consideration that every citizen expects from his fellow man and the Home Office."[26]

Henderson's was one of several cases that made officials in the Home Office begin questioning where the line between illegitimate and permissible opiate use lay. A group of doctors in Glasgow, for example, treated a patient similar to Henderson who had been taking morphine for twenty years by giving him prescriptions so that he could continue to work. In Leicestershire, authorities faced a difficult dilemma when they discovered a pharmacist who continually provided laudanum to an addict because he felt she would be "unable to do without it." In another case, a doctor continued prescribing morphine to a patient who had been addicted for twenty-five years, since this was the only way to keep him in treatment. While officials in the Home Office believed these were violations of the law, not forms of treatment, doctors maintained that, medically, they were acting in the best interests of their patients. "It is such a great pity," lamented one doctor in correspondence with the Home Office, "that there is no provision in the [Dangerous Drugs] Act for such existing cases." In all of these instances, the addicts in question seemed to be productive members of society who did no harm, so authorities needed to consider what good would come from denying them the small doses of drugs they needed to function.[27]

Toward a Resolution: The Rolleston Committee

Drug users and medical practitioners were not the only ones who were unhappy with the ambiguities in the existing regulations. Within a few years of passage of the Dangerous Drugs Act, officials in the Home Office grew frustrated with cases in which doctors and pharmacists seemed to be abusing their right to prescribe and dispense opiates, and they feared that many were feeding the appetites of addicts or, even worse, diverting supplies to the black market. In one case, for example, the Home Office learned of a London doctor who gave morphine prescriptions "to practically any person who asks for them," but it was unable to prosecute him because he was doing so under the guise of medical treatment. Some "draft doctors," as they were called, would prescribe drugs to addicts they had never met, and without consulting with their treating physician. One particularly notorious doctor wrote more than five hundred prescriptions for controlled substances for patients all over the country in a nine-month span in 1923 and 1924, claiming that the drugs were needed to help return "displaced internal organs" to their proper places in the body. Although the case "savoured of quackery," as the Home Office explained in a summary report, "it was very difficult, if not impossible to discover any charge which could be made against him in court," because the doctor claimed to be prescribing drugs as part of medical treatment for physical disorders.[28]

By 1923, officials in the Home Office had grown exasperated by how difficult it was to crack down on the "draft doctors," and it wanted to tweak the law so that they could be brought to justice. Furthermore, the Home Office was unclear on the question of whether the long-term provision of opiates to confirmed addicts such as Henderson was therapeutically legitimate or simply drug dealing under cover of a medical license. In January, Home Office officials wrote to the Minister of Health asking for his opinion on these cases. From the perspective of law enforcement officials with little medical training, the two issues were intertwined: it was difficult to determine whether opiates were being dispensed "properly" to alleviate suffering for individuals with medical conditions or were "being prescribed in order to enable [addicts] to gratify their addiction and without any design or hope of effecting a cure."

The answer was largely dependent on where the line between using drugs to help cure addiction and the mere satisfaction of addicts' appetites lay. If drugs were being prescribed as part of a plan to wean addicts off opiates, the Home Office maintained, the practice should be considered acceptable, but when the prescriptions continued for an indefinite period, it should not. The Home Office hoped, with the assistance of experts from the Ministry of Health, to lay down a set of regulations to this effect, similar to rules recently instituted in South Africa.[29]

According to the South African regulations, doctors were allowed to prescribe opiates for the treatment of disease or to alleviate pain from organic disease, but "the satisfaction or relief of a habit or craving for the drug" did not qualify as acceptable medical justification for dispensing opiates. Furthermore, unless the drugs were administered or used in a medical setting, the South African rules made the ordering, prescribing, selling, supplying, administering, and using of opiates illegal. These regulations, British Home Office officials believed, could be a "useful" model for determining whether opiate prescriptions were legitimate, and they hoped to implement a similar set of guidelines for Britain. Given that the question touched on a clearly medical issue—that of defining where pain relief ended and the facilitation of addiction began—the Home Office wanted input from the Ministry of Health as it drafted the new regulations.[30]

Health officials agreed with the Home Office that "the mere satisfying of the craving of a drug addict without further object" of a cure, even by a doctor, should be considered little more than "illicit dealing" through a prescription pad. If doctors could prescribe drugs simply to satisfy addicts' appetites, "the [Dangerous Drugs] Act could only be regarded as a measure designed to enable the addict to obtain his drugs legally and easily, which is absurd." At the same time, Health Ministry officials firmly maintained that drug addiction was indeed "a disease," a "pathological state calling urgently for treatment and relief." But unless addicts could be cared for in an institutional environment where they would be closely monitored, part of any "treatment and relief" would have to involve the administration of drugs on a gradually tapering schedule. Opiates, therefore, were not just the root of the problem but, in many cases, part of the solution as well. Though this did not mean that doctors should be free to become "purveyor[s] of drugs for all," it did mean that they should be allowed to dispense the drugs as part of a legitimate medical intervention for the treatment of diseases—including addiction itself.[31]

Consequently, Health officials maintained, "the physician must be allowed to use his discretion in the matter of dose of the drug to be employed and other details." It was not, as the Home Office had suggested, up to the state to dictate the when, where, and how of administration of opiates to addicts, as was the case in South Africa; these were clinical questions and, as such, were to be left to clinical discretion. Provided that "the relation of the addict and the doctor is that of patient and physician," and both parties agreed that addiction was a disease that needed to be cured, it would be difficult to set hard-and-fast rules dictating when opiate prescriptions were "legitimate" and when they were not. It was impossible, they concluded, to lay down an authoritative statement or set of rules governing the prescription of controlled substances in the way the Home Office had hoped. "The question," as one Health official summed up in August 1924, "can be determined only by reference to the facts of the particular case."[32]

Some Ministry of Health officials also maintained that there were indeed some cases, such as Henderson's, where a cure was "impracticable," and the best a doctor could hope for was "alleviation of the condition." This raised the question of whether, in such cases, addicts could be perpetually prescribed drugs within a "therapeutic" doctor-patient framework. Again, it was difficult to say, since the intractability of an addiction could only be determined based on the specifics of each case, and the imprecision of medical criteria could not always correspond to fixed regulations laid out by law. Thus any "theoretical criterion," one Health official warned, would become difficult to apply "in practice." If policies governing the prescription of opiates were to be formulated, they would have to be written in a way that allowed the treating physician, not a civil servant, to determine which prescriptions of opiates were permissible and which were not.[33]

When their initial inquiries to the Ministry of Health showed that the issue was more complex than they first imagined, officials in the Home Office realized that more research was needed. As historian Virginia Berridge explains, the Home Office started "seeking medical advice" on how to devise a scheme for regulating opiate prescriptions that was both legally and clinically sound. With the aim of drafting a more definitive policy, the Home Office asked the Minister of Health to appoint a committee to study the questions surrounding opiate prescriptions, and the group began work in September 1924. The Departmental Committee on Morphine and Heroin Addiction, formed that autumn, was headed by Sir Humphrey Rolleston of the Royal College of Physicians; it also included prominent representatives from the Home Office, the

British Medical Association, and the Ministry of Health. To inform policy, the Home Office wanted the Rolleston Committee to determine, based on expert medical opinion, where the limits of legitimately prescribing opiates lay, and it set out several other questions for the committee to consider. Were there indeed circumstances in which supplying opiates to addicts could be "medically advisable?" Was a doctor who dispensed opiates to addicts undertaking a "legitimate medical treatment," or was he simply "acceding" to the appetites of his patient? Were there really addicts, such as Henderson, for whom complete discontinuance of the drug habit was unattainable, or was it preferable for physicians to always aim to make their addicted patients drug-free? And if maintenance treatments were advisable, what rules or guidelines should practitioners follow when treating their addicted patients?[34]

In twenty-three meetings in which it heard testimony from thirty-four witnesses (twenty-four of whom were medical professionals), the Rolleston Committee worked for more than a year to answer the Home Office's questions before finally publishing its findings in 1926. Most of the witnesses before the committee maintained, along with the medical wisdom of the day, that addicts, if at all possible, should be treated in an inpatient setting where physicians could monitor their health and ensure their compliance. If the state were to demand the complete withdrawal of opiates from all patients, it would need to provide for establishment of such institutional resources. But given the lack of such treatment facilities in the 1920s, such a policy would be "impracticable," and there were going to be "persons in whom a complete cure cannot be expected." Furthermore, even though the state could have, in theory, invested the time and resources to create adequate space for institutional treatment, it was probably not worth the effort. Opiate addiction, the doctors testified, was rare in Britain: some witnesses at the committee hearings reported treating just two or three addicts in the previous twenty years. The small numbers of addicts who needed care was not enough to justify the construction or expansion of treatment institutions.[35]

Practical limitations and the small scope of the problem were not the only considerations that led the Rolleston Committee to conclude that maintenance treatment was acceptable in some circumstances. Many doctors who testified before the committee reported that there were some opiate addicts for whom complete withdrawal was too difficult to justify the undertaking. For such individuals, total abstention could come only at great cost, not just because of the time and resources needed for treatment; there was also no guarantee that addicts would be healthy enough to be productive and self-sufficient once drug-free. "Experience showed," the committee wrote in its summary of expert testimony, "that a certain minimum dose of the drug was necessary to enable the patients to lead useful and relatively normal lives, and that if deprived of this non-progressive dose they became incapable of work." Thus it was of greater benefit both to the addict and to society as a whole to allow "regularly the minimum dose which has been found necessary, either in order to avoid serious withdrawal symptoms, or to keep the patient in a condition in which he can lead a useful life." Maintenance treatment, in other words, was the best course for some addicts, and there was little to be gained—for the patient or for the community—if they were compelled to give up their drugs.[36]

To define what the limits of acceptable maintenance treatment were, the Rolleston Committee laid out guidelines for treatment in "apparently incurable cases" in which the "continuous administration of the drug indefinitely" was justified. Doctors needed to see patients at least once a week to observe their condition and reevaluate whether their addiction really was incurable. When dispensing drugs or writing prescriptions, doctors were advised not to give more than a week's worth at a time—a precaution that would both keep the addict coming back for treatment regularly and ensure that no addicts could stockpile and start selling drugs. If an addict was traveling, doctors could provide more than a week's worth of drugs, but only if they instructed the patient to check in with another practitioner. Even though the addict could continue to procure drugs legally under these guidelines, drug taking was still to be subject to tight regulations and medical surveillance, making it, as historian Alex Mold concludes, still strictly controlled, but now "through treatment."[37]

Though not written as an official policy document, the Rolleston Committee's findings proved influential, and the Home Office implemented its recommendations in a new set of regulations released shortly thereafter. Medical inspectors from the Ministry of Health and medical organizations exerted pressure on doctors who did not follow Home Office guidelines to change their prescribing practices. Under the regulations, a medical tribunal was authorized to discipline doctors who disregarded the committee's recommendations. Most doctors, however, did follow the guidelines, and no such tribunal convened during the interwar period. The flexibility of the recommendations was good not only for doctors but also, according to some, for addicts themselves. As G. Laughton Scott of the London Neurological Clinic explained in 1930, for individuals whose addictions had not affected their social lives or careers as long as they were able to take the drugs, there would now be places where they could legally procure opiates. Without these options, he believed, many addicts would have wound up in desperate situations and needed to start forging prescriptions, stealing, or going to street dealers to get the drugs they needed.[38]

Thus, by the end of the 1920s, Britain had developed what some commentators have termed the "British System" to treat cases of addiction that had no deleterious social side effects, while maintaining tight controls over opiates. As many historians have pointed out, a major reason that the Rolleston Committee came to the conclusions it did—and the Home Office was willing to accept them—was the relatively small number of addicts in Britain at this time. As Berridge says, British tolerance of maintenance treatment in cases of iatrogenic addiction was "the *result* rather than the *cause* of the low numbers of addicts in Britain and their middle-class profile." As the committee noted in its final report, the administration of nontapering maintenance doses of opiates was not the ideal; proper inpatient treatment was still the preferable course of action for treating addiction. Yet, according to the witnesses heard by the Rolleston Committee, opiate addiction was rare and, in some cases, socially unproblematic. Many of the addicts doctors saw were still leading "useful" lives in spite of their drug taking. So, while not preferable and only to be used as a last resort after other options had been exhausted, maintenance was permissible and would not cause significant social disruption if proper controls over drug supplies were maintained.[39]

The considerations that led to creation of the Rolleston Committee and publication of its findings revealed an understanding of opiate addiction as a disease that was not only rare but usually socially benign as well. As the productivity, self-sufficiency, and upstanding behavior of Thomas Henderson and others showed, opiate habits and those afflicted by them were not necessarily threatening. More importantly, the doctors who testified before the Rolleston Committee revealed that addiction and good citizenship were not necessarily mutually exclusive if drug use could help—rather than hinder—individuals as they carried out their roles as citizens in the community. Developments in British addiction policy during the 1920s, though definitely framed by the minimal numbers and generally middle-class character of most addicts, also revealed a British understanding of opiate addiction as a disorder that was not necessarily socially menacing. Because of who was becoming addicted, the number of addicts, and the fact that opiate use and addiction were not always irreconcilable with good citizenship, it was possible for some British addicts to exist among, rather than in isolation from, their fellow countrymen.

The Plight of Opiate Addicts in France

Given that the French and British drug control regimes initially placed similar limitations on doctors' freedom to prescribe opiates, many of the same questions and problems that emerged in Britain also surfaced in France in the 1920s. Unlike their British counterparts, however, the French did not launch any investigations on how to resolve the dilemmas posed by drug control, nor did they allow the maintenance treatment of opiate addicts. Thus, in France, much more than in Britain, all addicts remained in a state of medico-legal limbo throughout the interwar period.

Options for Addicts

Even before the institution of tighter drug controls in the early decades of the twentieth century, the treatment options available to opiate addicts were narrower in France than they were in Britain. In the late nineteenth century, the main reason for this was not just different attitudes concerning opiate use but a lack of institutional care available to other French victims of substance abuse disorders—alcoholics. The landmark French legislation dealing with alcoholism, the 1873 law on public drunkenness, focused more on the preservation of order than the creation of treatment facilities or programs for alcoholics. The legislation's only health-related provisions called for the creation of temperance societies to spread knowledge about alcoholism and its dangers. The French temperance movement, however, was generally still-born, since the Chamber of Deputies did not allot the funding the movement needed. In the late nineteenth century, some in the French medical community began advocating for the construction of public asylums to treat alcoholism, but their efforts yielded no results. In 1895, the Conseil général of the Seine approved plans for such an institution in the capital, but it was never built. The only way that alcoholics who were not wealthy enough to finance their own treatment in a private institution could receive help was to deteriorate to the point where they could be confined in an asylum. Some mental health institutions, such as the Ville-Évrard and Villejuif asylums in the Paris region, had special sections set aside for the treatment of alcoholics, but they provided little care specifically addressing drinking problems. Furthermore, since the rules governing confinement in these institutions were dictated by the 1838 law on the treatment of the insane, alcoholics could stay there only until the acute symptoms of alcohol withdrawal subsided; once they had dried out and no longer displayed any behavioral disturbance, most alcoholics were discharged—only to return when they began drinking and acting out anew.[40]

For opiate addicts who were institutionalized, the process was similar. Many addicts were put into asylums after they were arrested for behaving erratically when under the influence or for stealing to get money to feed their habits. Cases mentioned in contemporary medical literature show that most addicts were not given opiates when institutionalized, but were simply sedated to help them get through the pains of withdrawal. Yet, as with alcoholics, once opiate addicts no longer seemed violent or delirious, they were deemed "cured" of their drug-induced insanity and released without any aftercare. Consequently, it was not rare for addicts to regularly endure painful, sudden withdrawal processes in asylums, only to relapse and wind up back in an institution shortly after their release. Those who had greater means at their disposal could get help before they ran into legal or psychiatric troubles by checking into a private treatment facility.[41]

Even before the controls over opiates became tighter in the twentieth century, some medical practitioners began advocating for the creation of "special hospitals" for the treatment of opiate addicts deemed "incapable of curing themselves." At a meeting of the Société de médecine légale in 1899, practitioners suggested adding a provision to the 1838 law that would allow morphine addicts to be designated legally insane. By the time the government was planning the specifics of the 1916 law, some legislators were also won over to the cause. Socialist deputy Jean Colly, for example, recognized in 1913 that the current provisions for the treatment of addicts were sorely lacking. Putting addicts in asylums that were ill-equipped to treat them, he told the Chamber of Deputies, would be both an "inhumane" and "completely inefficient process," one that could cause much pain and suffering for the addict while doing little to cure him. To enable addicts to get more substantive help, he suggested adding a provision to the law that would call for the construction of facilities to treat drug addiction. Ultimately, however, the control regime that took effect in 1916 made no such allowances, and its implementation focused on suppressing the illicit traffic and controlling the flow of opiates. Through provisions restricting possession and use, the law also wound up leading to the prosecution of many addicts.[42]

The decreased availability of drugs that resulted from the 1908 decree and the 1916 law indirectly forced many addicts into treatment or simply forced them to quit using drugs on their own, as it became exceedingly difficult to maintain opiate habits without running afoul of the law. As the restrictions took hold, many addicts made do by using fraud to obtain opiates from otherwise legal sources: presenting pharmacists with fake or stolen prescriptions, trying to reuse old ones, or tampering with the scripts from doctors to increase the amounts prescribed. According to historian Emmanuelle Retaillaud-Bajac, some sixty-five percent of users who faced legal proceedings in the interwar period were charged with such offenses. As pharmacists became more vigilant in refusing to honor false prescriptions in the 1920s, many addicts were compelled to turn to less savory sources. Some went directly to street dealers or frequented establishments that were well-known hotspots for the drug traffic. The more well-off were able to reduce the personal risks involved by sending servants or housekeepers to procure the drugs for them. When availability from legal sources waned by the 1920s, however, fewer and fewer Frenchmen could afford to keep up with soaring black market prices, making the opiate habit not just unhealthy and dangerous but prohibitively expensive. This probably led many addicts who could not or would not quit to turn to petty crime or dealing.[43]

For addicts of more substantial means, treatment in a private facility remained the best option after the controls of 1908 and 1916 took effect. The Villa Montsouris in the Paris region, the neurological clinic at Saint-Cloud, the private Sanatorium de la Rougière in Marseille, and other clinics in Bordeaux, Lyon, and southern France continued to accept private clients for addiction treatment. Since they had to provide quick results for customers who paid hefty prices for their services, most private clinics strove to achieve relatively rapid cures, weaning patients off drugs within a week and following up with a convalescence and recovery period. In 1922, a more affordable option became available with the opening of the Hôpital Henri-Rousselle on the campus of Sainte Anne hospital in Paris. Unlike inpatient institutions, Henri-Rousselle was an "open hospital"—one that did not have closed wards—and was designed to provide an alternative for patients who needed help with psychiatric problems but were not impaired enough to be interned in an asylum. Many opiate addicts who recognized that they needed help matched this demographic perfectly. By the mid-1920s, doctors across France began referring their addicted patients to Henri-Rousselle, and by 1930 it had become one of the most prominent addiction treatment centers in the country.

On average, the center received forty addicts per year in the 1920s, and about double that in the 1930s. Given the lower cost of treatment at Henri-Rouselle, its practitioners carried out their cures more gradually, usually taking about forty days to wean patients off drugs with a regimen of sedatives, baths, and psychotherapy.[44]

Despite the relatively modest costs at the Hôpital Henri-Rousselle, the price was still too high for many. The reason was not just the cost of treatment but the attendant and potentially devastating financial cost of missing several weeks or months of work. The onset of the economic depression in 1929 exacerbated this problem, and as addiction experts observed throughout the early 1930s, the financial climate made the combination of paying for treatment and losing salary too burdensome for most. Even in the best of times, practitioners noted, patients had to use an entire year's vacation time to undergo the briefest of cures, and with the economic crisis, time to withdraw and properly rehabilitate was a luxury that few individuals had. For financial reasons, Dupouy lamented in 1935, most addicts would "insist that their cure be done as quickly as possible," meaning that the preferred, more gradual methods had to be abandoned. To help make more rapid detoxifications successful—or at least bearable—practitioners increasingly turned to substitution treatments, sedatives, and unorthodox methods such as the injection of vegetable lipids. The shift toward more drastic withdrawal methods, though understood to be a practical necessity to accommodate patients' financial needs, was, in the view of many doctors, an unfortunate compromise. Nonetheless, it was a concession that needed to be made if addicts were to get any treatment at all, though it meant treatment was less likely to be effective or enduring. Consequently, the "cures" given at Henri-Rousselle rarely worked. Significant numbers of patients who entered the program either quit before completing it or relapsed after leaving.[45]

For addicts who either could not afford or chose not to use private treatment centers, the prognosis was not particularly good either. As in the nineteenth century, many did not enter treatment until they began having legal, psychiatric, or medical troubles that put them in a prison, asylum, or hospital. These institutions, though they took addicts in, did not make many special provisions for their care. Neither the prisons nor the asylums in France had procedures for weaning incarcerated addicts off drugs, and often the only treatments inmates received were drugs or restraints designed to control, quiet, and sedate them. Addicts who were arrested and began suffering from withdrawal while at the police station were usually taken to a separate depot designated for mentally ill individuals in police custody. Once arrested addicts made it through the worst of their withdrawal symptoms and their behavior stabilized, they were reintegrated into the general incarcerated population. Many who were detoxified in this manner relapsed soon after they regained their freedom. For those in asylums, there was little difference between the treatment received in 1930 and fifty years earlier: addicts were held as "insane" while suffering through withdrawal, and then released once they no longer seemed a threat to themselves or others. In some extreme cases, addicts were taken to an infirmary or medical specialist, but they were rarely given drugs to help ease the pains of withdrawal. Usually they were just monitored to ensure that they were not in any mortal danger. In public hospitals, the procedure was also relatively hands-off, as addicts were sometimes given bromides to calm them down but were otherwise subjected to immediate or very rapid withdrawal. Given the lack of aftercare, they left the hospital weaned but weak, and it was not uncommon for released addicts to return to drugs as soon as they got the chance.[46]

Early Critiques of the Drug Control Regime

As in Britain before the Rolleston Committee's report, addicts in France had few good options: they could bear the financial burden of paying for treatment, try to break free of opiates on their own, or resort to the black market for provisions. By the mid-1920s, some critics in France took notice of the situation the control regime had created for addicts and began calling for change. The most prominent of these critics was Surrealist writer Antonin Artaud, who had become addicted to opiates after first using them for medical reasons.

Artaud did not mince words in his assault against what he termed the "stupid laws," calling the advocates of narcotics control a collection of "medical asses, druggists of dung, dishonest judges," and "pedantic inspectors." He put forward a powerful critique of the French drug control regime on several fronts. Pointing out that many addicts turned to drugs for pain relief rather than psychotropic pleasure, he argued that the state should not dictate how those suffering from pain of any sort could ease their discomfort. The government, he claimed, had no authority to place limits on the individual's right to relieve his own suffering, be it of a "mental, medical, physiological, logical, or pharmaceutical nature." Such attempts accomplished little except to unjustly persecute people who were already subject to great discomfort through little fault of their own. Those who were sick should have what Artaud bluntly termed their "inalienable right . . . [to] be left the hell alone" respected. He also made a case on a practical front, arguing that prohibitions of any sort were bound to fail. Citing the shortcomings of alcohol prohibition in the United States, he maintained that control efforts encouraged the growth of the black market but did not curb use. "*The law of the forbidden fruit*," as he called it, actually worsened the problem: besides creating an incentive for criminals to start peddling drugs, he claimed, it "multiplies the curiosity" about drugs instead of turning people against them.[47]

An even more powerful critique was Artaud's assertion that any attempt to curb what opiates represented in the French imagination—physical and moral decline—was bound to fail. Critics attacked opiates, he maintained, not simply because the drugs were addictive, but because they seemingly posed a "danger . . . to the ensemble of society." But in spite of the state's best efforts, he maintained, there would always be individuals who were "congenitally maladapted" to a healthy and social existence. Some suffered from "cancers of body," while others were tainted by "cancers of . . . soul" that would draw them to poisons that were not only physical, but also spiritual and social. "Whatever it is . . . the poison of morphine, the poison of reading, the poison of isolation . . . the poison of alcohol, the poison of tobacco, the poison of anti-sociability," individuals who were wanting in physiological and emotional health would always be drawn to something that could help them cope and escape. Such "poisons," both chemical and spiritual, were not causes of physical or moral sickness, according to Artaud, but rather consequences of them. To focus on the symptoms of the disease rather than the disease itself, he claimed, was an act of folly that would ultimately accomplish nothing. "Deny them a means of madness," he warned, and "they will invent ten thousand others." Artaud emphasized that this applied both to substances that had physiological effects and those that had spiritual and social ones. "Nature itself is anti-social in spirit," and in trying to deny the outcasts of society their drugs of choice, "the social body [was] reacting against the *natural* slope of humanity." There would always be "men who always lose themselves," so there would always be some who would find new "means of getting lost," be they opiates or some other shortcut to distraction and isolation. Ultimately, therefore, the state had no business trying to dictate what means of "getting lost" men used: this was a fundamentally individual question that "*has nothing to do with society.*" Thus the drug control enterprise was flawed on two levels: on a practical level, it failed to diminish the availability or appeal of opiates, and since indulgence in psychoactive drugs was a fundamentally individual practice, society should not try to, and could not, stop it.[48]

Though not as philosophical or bombastic as Artaud, doctors also began to recognize the shortcomings of the strict French control regime by the mid-1920s. Some feared being prosecuted for doling out opiates too easily, and to avoid legal trouble they stopped giving them to any patients—even for the alleviation of physical pain caused by organic disease. Other doctors, however, still viewed opiates as valuable medicines and continued to dispense them for a variety of ills. In spite of increased knowledge about the dangers associated with opiate prescribing, this sometimes led to new cases of addiction. In the 1920s, French patients receiving morphine or heroin for neuralgia, stomach pains, insomnia, and asthma developed iatrogenic addictions and wound up in treatment, and others who took opiates under medical supervision to soothe depression or nervous disorders also had a hard time controlling their drug use. In these situations, particularly those involving patients with mental illness, doctors preferred to avoid the complications associated with withdrawal and suggested that rather than attempting a "radical cure" for addiction, it would be more prudent to reduce patients' treatments from "calming doses" to smaller, "medical" ones. Due to the nature and severity of the problems that led certain patients to start taking opiates, as well as the intensity of their addictions, doctors believed that, for some addicts, any attempts at withdrawal would be "vain and sterile."[49]

Even if clinically inadvisable, however, it was legally imperative for clinicians to wean their patients off opiates once their medical and psychological crises had passed. According to the 1916 law, opiates could be given only for "medical purposes," and the maximum duration of prescriptions was seven days. In practice, doctors generally had the latitude to prescribe opiates as they saw fit for the treatment of physical disease, but French prosecutors and judges stood firm in their belief that addiction itself was not a medical condition that justified continued opiate use. If the drugs were to be legitimately prescribed as part of addiction treatment, the authorities maintained, this needed to be done in progressively tapering doses, with the ultimate goal of helping addicts become abstinent. To do otherwise, particularly for outpatients, was considered akin to "giving a bottle of wine to a drunkard with the recommendation that he use it with discretion throughout the week." But for the physicians providing treatment, the demarcation of "legitimate" and "illegitimate" prescribing was often unclear. In cases of iatrogenic addiction, this line was particularly hard to discern, since addicts' suffering when they reduced their drug intake could be interpreted either as resulting from addiction-related withdrawal symptoms or as original pains reemerging once opiates' palliative effects wore off. Thus the questions brought up by opiate prescriptions, and the matter of when to limit them, posed challenges that were both clinical and legal. Often, it was only clinical opinion and discretion that could determine the boundary between responsible medical opiate prescriptions and the reckless doling out of drugs to addicts. As a commentator writing in the *Annales Médico-psychologiques* succinctly summed up in 1923, "it is easier to discriminate between the two on paper than it is in practice."[50]

Anxious about the implications of the 1916 law for their medical practice, and indignant about the sometimes harsh responses of the justice system to the suffering of addicts in withdrawal, many doctors began to argue that France's penal regime needed a sanitary corollary. The reason that addiction continued to be a problem after 1916, doctor Jules Ghelerter of Henri-Rousselle argued, was the erroneous assumption that "the addict is a delinquent, which he is sometimes, but not . . . a patient, which he is always." Addicts were "patients in every sense of the word, and need to be treated as such," with attention focused on minimizing their physical and psychological pain, not on punishing them for legal or moral transgressions. Yet more often than not, Ghelerter concluded, the authorities tended to treat addiction as more of a vice than a disease or disorder. Georges Boussange, a student at the Paris Faculty of Medicine, agreed, arguing that the penalties for violating drug laws in France were "too rigorous." By virtue of his disease, the addict was already subjected to "the horrible torture that results from immediate withdrawal" when unable to procure drugs. Since addicts had already suffered plenty from "rigors that are unforeseen and not imposed by the law," he concluded, the state did not need to take further punitive action against them. As the newspaper *Paris Soir* summed up in 1927, there was a growing sense in the medical community that "the current legislation is too rigid, or at least interpreted too rigidly."[51]

Toward a Resolution?

Despite these critiques, the French undertook no formal investigation into the ambiguities and problems brought about by the control regime, as the British had done. This is not to say that many of the same questions that provoked the creation of the Rolleston Committee in Britain did not exist in

France—they certainly did. As in Britain, addiction treatment specialists in France recognized that some patients, given the practical restraints and the difficulty of finding a lasting cure, could be classified as "incurable" addicts. And as in Britain, French doctors recognized that many of these patients could still function as contributing members of society in spite of their addictions without displaying any overtly antisocial or dangerous behaviors. Doctors at the Hôpital Henri-Rousselle, in particular, found that many addicted patients were petit bourgeois and workers whose drug habits never interfered with their everyday activities; they were able to attend work regularly and remain "sober and prudent," even while injecting opiates daily. For these addicts—who could be considered the French counterparts of Thomas Henderson—doctors were inclined to provide indefinite maintenance treatments, if given the choice.[52]

But whereas in Britain, the Home Office took notice of these dilemmas and solicited medical advice on how to deal with them, the French authorities did not. Usually it was left to the courts to decide when opiate prescriptions were within the bounds of the 1916 law, thus creating situations where judges— trained in the law, but not in medicine—were forced to make rulings based on their limited clinical knowledge or by relying on the clinical knowledge of others. Generally, judges were conservative in their interpretation of the law and ruled against doctors accused of prescribing nontapering maintenance doses of opiates. In 1923, for example, a group of forty-nine addicts, along with six pharmacists and eight doctors, were charged with violating the 1916 law by undertaking detoxification cures with opiate prescriptions that exceeded seven days. As an expert witness from the Academy of Medicine told the court, it was often difficult to determine where the line between treating and facilitating addiction lay, since the medical community was still divided on what the best detoxification practices were. To be considered part of a treatment for addiction, he testified, doses needed to be "progressively lessening . . . until a complete cure" was reached. If doctors were to prescribe opiates in the course of addiction treatment, he believed, they should provide only tapering doses, with the ultimate goal of helping patients become drug-free. He cautioned, however, that it was not necessarily up to the courts to decide whether the case represented a violation of the doctors' professional responsibility. Nonetheless, the court rendered a guilty verdict for some of the doctors, sentencing three of them to between two and three months in prison, and punishing one who had a history of offenses against the 1916 law with two years imprisonment and a heavy fine.[53]

Subsequent cases further established that even though the question of whether prescriptions were designed to cure or facilitate addiction was a medical matter, judges—not doctors—had the prerogative to make the final decision. Though never written into law, by the 1930s, the rulings of most judges made clear that opiates could still be used to treat other illnesses, but maintenance prescribing was not allowed in addiction treatment. As one judge found, "while generally the doctor definitely has the right—and often the duty—to spare certain patients, such as cancer patients . . . useless suffering, and he can undertake some detoxification cures, he cannot facilitate drug use" by prescribing opiates to addicts, unless in diminishing doses with a clear aim of abstinence. Although judges would grant clemency in certain situations, particularly those involving relief for veterans who developed iatrogenic addiction during World War I, law enforcement's general tendency was to investigate and crack down on doctors, pharmacists, and addicts who seemed to take advantage of the 1916 law's provisions that allowed medical prescription of opiates.[54]

By the mid-1930s, the consequences of the law's tight interpretation became clear to most doctors in France. Increased awareness of both the clinical and legal risks of using opiates to treat discomfort led practitioners to provide the drugs sparingly, with experts advising that morphine should not even be used to relieve the pain of amputees, and that it should be reserved for use as a palliative for the extreme suffering of cancer patients. For the treatment of addiction, many medical providers recognized that application of the law forbade maintenance, and they were troubled by the ethical and legal conundrum that the control regime placed on doctors treating addicts. "It is undeniable that the legislator did not want to create these conflicts," pharmacist G. L. Tourade wrote in a 1934 critique of the way the 1916 law was functioning. The legislator "only wanted . . . the [addicted] patient . . . [to] remain under the constant surveillance of the doctor," not necessarily to be denied the drugs needed to lead a normal and productive life.[55]

Some practitioners, despite the legal risks, agreed with Tourade and took risks to treat their addicted patients as they saw fit, even if it meant taking some creative license with the regulations. One doctor, for example, tried to make it easier for his addicted patients to avoid legal trouble at the pharmacy by giving them certificates verifying that they needed the drugs as part of a "necessary treatment." Such documents had no legal value, but they did increase the chance that pharmacists would be sympathetic and perhaps turn a blind eye when doctors prescribed more than a week's worth of opiates on a nontapering schedule. According to some addicts, doctors and pharmacists who were willing to do this—and resort to other tricks to help them procure the drugs they needed—were not difficult to find if one knew where to go. Nonetheless, many practitioners did not want to be stuck in a situation where they had to choose between the restrictions imposed by judges and the needs of their patients. Some doctors, though understanding the need for strict controls, recommended creating special provisions in the law for patients with intractable cases of addiction. One suggested issuing special cards to medically certified morphine addicts that gave them the right to possess the drug, thus safeguarding both the doctor's professional prerogative and the patient's well-being.[56]

While some French doctors advocated for maintenance treatment of addiction along the British model, others thought the French law could be improved by making it tougher, not easier, for addicts to continue living with their drug habits in the community. Citing laws passed in Brazil and Belgium in the early 1920s and policies put in place in the United States at the end of the decade, many addiction specialists wanted the state to "impose treatment" on opiate addicts. Compulsion would be good, they suggested, not only because it could help force people into the care of a physician, but because—as was the case in prisons and asylums—it could make cures swift and decisive, with little regard for patients' fears or desires to continue using. Ghelerter of Henri-Rouselle, for example, estimated that about one-fourth of the addicts whom he saw quit treatment against medical advice, and he suggested that an element of compulsion could help improve the chance of a cure. If supervised weaning followed by aftercare could be made "obligatory" and "mandated to be finished" in an institutional setting, doctors could help addicts defeat their afflictions instead of simply trying to convince patients to overcome them. "The French law is insufficient in this respect," Ghelerter charged, suggesting that the law be amended to force addicts to "complete" their cures, and that addicts who relapsed be interned in a hospital setting to be "reeducated" once and for all. Others echoed the call for mandatory institutional care; if the state did not want to finance the construction of facilities to care for addicts, they claimed, it should at least set aside special wards for detoxification and rehabilitation within the public hospital and asylum systems.[57]

Calls for a sanitary corollary to France's drug laws, even one that had elements of coercion, had little effect in the interwar period. An approach that did gain traction was the other action suggested by the French medical community: to crack down on the illicit traffic. Many French doctors advocated a system that would make greater distinctions between the victims of addiction and drug pushers, arguing that the most effective—and just—enforcement measures would prioritize the prevention and punishment of illicit dealing, while showing greater clemency for drug users. Thus, a renewed focus on suppressing drug supplies would make France's struggle against addiction both more just and more effective. In 1921, Marcel Briand and Paul Cazeneuve of the Ligue d'hygiène mentale argued that more enthusiastic enforcement measures, with rewards for police who made drug arrests and instructions for judges to mete out harsher sentences to drug law offenders, could choke off supply lines and stamp out the problems of drug use and addiction. According to some commentators, it was not just the organized trafficker or street corner dealer who needed to be more closely monitored, but also the complaisant doctor. Benjamin Logre, a physician with the Paris police department, suggested that judges should mercilessly punish doctors and pharmacists who dispensed opiates in violation of the law, and that they should automatically be sentenced to prison, no matter what the circumstances.[58]

By the early 1930s, though not responding to pleas for greater prescribing flexibility, the French government did start to heed calls for stricter controls. With a series of decrees and regulations issued between 1930 and 1934, it instituted more stringent regulations governing the labeling, manufacturing, distribution, production, and trafficking of substances covered by the 1916 law. In the summer of 1939, a decree aiming to increase the birthrate and protect the family integrated a new, more stringent set of punishments for individuals who violated the laws on drug trafficking, raising the maximum fine to a hundred and twenty thousand francs and the maximum prison term to five years. Thus, by the dawn of World War II, the French drug control regime had become, as historian Igor Charras concludes, "one of the most draconian in Europe" in controlling the flow of opiates, but among the least flexible in the treatment of addiction.[59]

Different Drug Problems, Different Drug Discourses, Different Drug Policies

Britain and France were faced with similar drug policy questions after tighter narcotics controls took effect. By the early 1920s, some doctors and opiate addicts in both countries voiced concerns that the control regime was not effectively checking the spread of addiction, and that it was morally and ethically wrong to deny an individual suffering from drug dependence the substances he needed to live an otherwise normal life. In both countries, critics suggested that the law should take into account that some addicts were "incurable" yet nonetheless productive members of society. Even though British and French citizens touched by the new controls shared some common concerns, the British authorities were willing to allow maintenance treatment in intractable cases of opiate addiction, but the French were not. Why this difference?

The development of policies—or lack thereof—to address opiate addiction in France was certainly influenced by understandings of how many Frenchmen, and what kind of Frenchmen, had opiate habits. Historians believe that the British addicted population was probably in the hundreds in the 1920s and 1930s; for the same period, Retaillaud-Bajac estimates between one and ten thousand regular users of controlled substances in France. Based on population statistics for 1931, this would mean that about 0.0015 percent of Britons (fewer than one in sixty-six thousand) were using illicit drugs regularly, compared with 0.012 percent of the French population (fewer than one in eight thousand). Roughly speaking, this would mean that regular use of controlled substances was approximately eight times as prevalent in France as in Britain. However, the higher prevalence of cocaine users in the French statistics probably makes the quantitative difference between Britain's and France's opiate-using populations less drastic. About 95 percent of the British addicts reported in these statistics were opiate users; a much more significant proportion of French than British drug users—at least, among those appearing in statistics garnered from historical judicial records—were cocaine users. Yet, even if we assume that only half of the French users who show up in these statistics were using opiates, that would still mean opiate use was four times as prevalent in France as in Britain. Limitations in historical documentation make it difficult to ascertain precise prevalence rates, but it is nonetheless clear that, numerically, opiate use and addiction were significantly more common in France than in Britain during the 1920s and 1930s.[60]

Not only was opiate use more common in France, but its character was more varied. In Britain, the lion's share of opiate users were iatrogenic addicts, generally middle class, and not considered socially disruptive or culturally deviant. France also had a large share of opiate users whose drug habits and general behavior did not cause much alarm or concern, but other types of opiate consumption were also prevalent in the interwar period. Opium smoking continued in the armed forces and among Asian immigrant populations, and morphine and heroin were used recreationally alongside cocaine in fashionable and artistic circles. A subculture focused on both trafficking and using opiates (often in conjunction with cocaine) was also beginning to emerge in France. Many French opiate users were turning to more powerful substances in the 1930s, as heroin began to displace morphine as their drug of choice. The opiate-using population in France was also younger than that in Britain; statistics show that French opiate users tended to be in their twenties or thirties, whereas in Britain they were middle-aged. Thus, just as in the late nineteenth century, nonmedical opiate use in France in the 1920s and 1930s was seemingly more widespread across the socioeconomic spectrum and more publicly visible and potentially socially disruptive than in Britain.[61]

Yet, beyond differences in how many people were using opiates, who they were, and which drugs they were using, there was another major divergence in the concerns about drug use in interwar Britain and France: its perceived social and political significance. Going back to the nineteenth century, British anti-narcotic nationalism had centered around fears that opiate use could decrease independence, self-sufficiency, and industriousness—characteristics essential to British understandings of citizenship. In 1916, when opiates began menacing the smooth functioning of British commerce, the state was driven to action and instituted tighter controls to ensure that the flow of drugs would not compromise the nation's trade interests. By the 1920s, however, the links between opiates, idleness, and commercial problems had declined. The Rolleston Committee's recognition of Britain's small, generally middle-class, and upstanding addicted population allayed fears about drug habits and what they could do to British society. Thus, much of the potential threat to the British way of life represented by opiates at the beginning of the twentieth century had been neutralized a few decades later. In fact, as the investigations by the Rolleston Committee showed, it was sometimes more practical to let addicts take opiates under medical supervision and lead productive lives than to deny them their drugs. Given that opiate use and good citizenship were not necessarily irreconcilable, allowing maintenance treatments so that addicts could continue to live as free and functioning members of the community was not inconsistent with but, in fact, supportive of the tenets of British citizenship. Thus the British approach to opiates and opiate addicts in the 1920s, though largely a product of the size and demographics of the opiate problem, was also colored by understandings of what opiate use meant and how opiate users could fit into the larger social and political whole.

In France, the anti-narcotic nationalism that had developed in the late nineteenth century proved more enduring. Regardless of their profile or their numbers, opiate users—by the very nature of their drug habits—were discursively framed as poor citizens: individuals who did not engage with the social whole, did not contribute to the common good of the nation, and often posed a threat due to their links with France's enemies. Even outside the colonial context and after the Armistice, opiate users were framed as threats to the nation. Within the political culture of the Third Republic, opiate habits were seen not only as distasteful but as threats to the social and political fabric that both bound and defined the French nation. Such thinking transcended the realm of discourse and became entrenched in policy as the nation's drug control apparatus took shape—allowing the state to effectively strip drug law offenders of their basic rights as citizens. Whereas, in Britain, initiatives to control the flow of opiates were largely a means to an end, in France, they were ends themselves, and the crusade against opiates and opiate users was not simply practical but also highly nationalistic and ideological.

In the French context, therefore, it is hardly surprising that despite the recognition that some addicts were upstanding members of their community, there was no move to address their health needs—either by allowing maintenance treatments (as the British did) or by providing institutional treatment with an eye to a cure (as did authorities in countries with more widespread drug problems, such as the United States).[62] The relative indifference to the suffering of addicts in France blended into a larger pattern of thinking about opiate users not just as victims of a disease or disorder but also as poor citizens. Thus, although the size and character of the French drug-using population certainly played a part in shaping the development of France's addiction policies, understandings of opiate use as irreconcilable with good citizenship also influenced French action—or inaction—when it came to addressing the problems that opiate use caused for the nation's addicted citizens.

Changes and Continuities

Since World War II, vestiges of British and French approaches toward narcotics control and addiction treatment dating from the 1910s and 1920s have remained intact, but attitudes have also evolved to meet the changing realities of each nation's drug problems. Two major developments led to shifts in drug policies: the dramatic increase of drug use among the young in the 1960s and the emergence of AIDS as a public health crisis, connected with drug use, in the 1980s. Yet the ideals of citizenship have remained consistent undercurrents in how drug use and the societal challenges it poses are understood, and they continue to influence how industrialized societies address the drug problem today.

The 1960s and 1970s: Tightening Controls

In Britain, the drug control and treatment regimes established after the Rolleston Committee remained in place for the next forty years. In the 1930s, some medical practitioners suggested that the government should pass legislation to impose mandatory treatment on addicts, but the Home Office found the suggestion politically untenable and allowed the system of limited maintenance prescribing to continue. By the early 1960s, however, it became clear that a handful of London doctors were taking advantage of the system by prescribing heroin too liberally. Furthermore, the number of addicts known to the Home Office, though small, began to grow, more than doubling between 1959 and 1965; most were heroin users. That the increase was most pronounced among the young, and among those who did not begin using drugs for therapeutic reasons, was of particular concern. Even more alarming was the emergence of a new "heroin subculture," generally composed of individuals of low socioeconomic status who engaged in petty crime and prostitution to support their drug habits.[1]

Taking notice of these developments, in 1965, a Ministry of Health interdepartmental committee on drug addiction declared addiction to be a "socially infectious" condition. To address the issue, the committee suggested establishing special treatment centers with the power to detain addicts, and making these centers the only places where prescriptions for heroin could be given. Not all of the recommendations were implemented, but a set of regulations that took effect in 1968 established protocols to prescribe heroin at specialized clinics, most of which were located in hospitals. In the 1970s, addiction treatment practices varied among practitioners, with some still prescribing heroin in maintenance doses and others using methadone substitution or other methods. Over the course of the decade, however, maintenance prescribing of heroin declined precipitously, with most providers shifting their patients to methadone, then working to gradually decrease the doses until individuals could become drug-free.[2]

Drug use also became a matter of increased public concern in France in the 1960s, particularly because of its association with the upheavals of May 1968. Largely in response to public fear of unruly, potentially revolutionary, and possibly intoxicated youths, in December 1970, the French government passed a new law to suppress drug use. The law took many steps to strengthen the repressive measures put in place at the beginning of the century: stiffening penalties, closing a loophole in the 1916 legislation by explicitly forbidding the use of controlled substances even in private, and making it an offense to "incite" drug use by framing it in a positive way. Just as the state had reacted swiftly and decisively in its fight against drugs in the World War I era, in the anti-drug campaign of the late 1960s and early 1970s, the government's response was, as Tim Boekhout van Solinge sums up, as "ferocious as if the very republic itself was at stake." While acknowledging that the new law entailed an exceptional violation of civil liberties, French policymakers concluded that such measures were justified because, as one lawmaker explained, "the drug user, by virtue of the vice that he adopted or was imposed on him, has lost a large portion of his right to liberty." The new restrictions were rigorous not only on paper but also in practice, as the number of drug arrests grew every year between 1970 and 2000. Even activities or behaviors only tangentially related to drug use, such as possessing syringes or wearing T-shirts with messages considered "incitements" to use drugs, led to some arrests under the 1970 legislation. As Alain Ehrenberg points out in his analysis of the 1970 law, conceptions of citizenship played a key role in shaping the French legislation and making its implementation so harsh. Drug use was seen as an affront to the norms of French republicanism, privileging private passions over civic engagement with the nation. Thus, just as in the 1910s and 1920s, ideas concerning drug use and its relationship to social cohesion and citizenship shaped France's responses to the drug problem in the 1960s and 1970s.[3]

Though oriented toward repressing and checking the spread of drug use, the 1970 law also had a therapeutic side, establishing procedures for courts to order addicts to be cured by the state instead of going to prison. French judges, however, used this sparingly, applying it in only about five percent of cases in which it could have been used as a sentencing alternative. Even for addicts who did find their way into treatment, the services they received rarely included maintenance or substitution treatments, as such services did in many other countries. Viewing addiction from a psychoanalytic perspective, most French practitioners in the 1970s considered addicts' troubles to be outgrowths of repressed childhood traumas or other causes rooted in the personality rather than in the body. Consequently, practitioners tended to treat addicts with psychological rather than physical tools, offering plenty of talk therapy but little in the way of pharmaceutical interventions to ease their suffering. When addicts did not respond to psychoanalytic treatment, providers generally viewed this as an indication of unwillingness to put forth the effort to get better, rather than a sign that other interventions were necessary.[4]

Some doctors did advocate for substitution treatment, but it took until the end of the 1970s for France to open methadone programs, and then only on a limited and experimental basis. Most providers remained critical of methadone, however, arguing that it did little except replace addicts' dependence on one drug (heroin) with dependence on another (methadone), and that medical interventions were of little use because they failed to address addiction's underlying psychological roots. In 1978, the French government's Pelletier report affirmed the addiction treatment community's anti-methadone consensus, and interventions aimed at helping addicts overcome their afflictions without the use of substitution drugs remained the norm in French treatment services. Furthermore, since the 1970 law banned "incitement" of drug use, it was difficult for providers to offer addicts advice on how to reduce the harms associated with addiction; by offering advice on how to inject safely, they risked violating the law. Thus abstinence, rather than harm reduction, became the goal of nearly all addiction treatment services in France during the 1970s and 1980s. Unlike in some other countries, where law-and-order measures coexisted in uneasy tension with treatment initiatives, the two acted in concert in France, pushing for a drug-free nation and a drug-free addiction treatment system. The alliance of law enforcement with the treatment community undergirded what Ehrenberg calls the "golden triangle" of French drug policy, with "abstinence as its foundation, detoxification as the goal for the user, and the eradication of drugs as the goal for society."[5]

Thus, even though the British stance on maintenance treatment became more restrictive in the 1960s, Britain was still much more willing than France to provide at least some sort of extended medical treatment to its addicted citizens. The consequences of the different approaches became clear by the mid-1970s and 1980s. The British used methadone widely in the treatment of addiction, but the French used it rarely, usually providing other, non-opiate drugs—such as clonidine or benzodiazepines—to ease addicts through the pains of withdrawal. Maintenance treatment was extremely rare in France, provided for fewer than one in eight hundred heroin addicts in the 1980s; in the United Kingdom it was used for somewhere between one in fifteen and one in forty addicts. Furthermore, methadone prescriptions were time-limited and under strict surveillance in France, whereas the British allowed indefinite methadone maintenance, and addicts could fill their prescriptions at pharmacies outside the drug clinics. Not only were significantly fewer French addicts receiving methadone than their British counterparts, but they also received much less of it, due to tighter dosing restrictions in France. Thus, although the British were becoming stricter in cutting back on heroin prescriptions and maintenance treatments, their addiction treatment practices remained quite liberal—in terms of what treatments they provided and on what scale—when compared with those of the French.[6]

The 1980s and 1990s: Addiction and AIDS

The second major shift in British and French drug policies after World War II came in the 1980s and 1990s with the AIDS crisis, particularly after the connection between intravenous drug use and the transmission of HIV became clear. From both a public health and a public policy perspective, checking the spread of AIDS was more important than pushing addicts to quit their drug habits, and most industrialized nations adopted harm-reduction policies in the late 1980s and early 1990s. Programs designed to prevent the sharing of needles, to persuade drug users to switch from injection to other routes of administration, and to educate the public about safe sex all emerged as policy responses to the threats posed by AIDS. Thus AIDS helped recast services for drug addicts as more than a set of narrowly tailored interventions designed to minimize drug use; it led to the integration of addiction treatment into broader initiatives focused on disease prevention and public health.[7]

In Britain, AIDS did not bring about dramatic or novel changes in drug policy as much as it reoriented policies toward harm reduction. In the mid-1980s, as British health officials recognized the danger of HIV spreading from injection-drug users to the general population, reducing the harms associated with drug use became a more pressing matter than trying to eliminate it altogether. In 1988, the Advisory Council on the Misuse of Drugs concluded that HIV posed a greater threat than drug use, and it called for changes in drug policy that would, above all else, minimize the risky behaviors that spread the disease. The principal aim of drug treatment policy and practice thus shifted toward encouraging drug users to move from injection to oral use, and abstinence was relegated to a secondary goal. Soon thereafter, the British began establishing more needle-exchange programs, and by 1990 some two hundred programs were up and running throughout the nation. Although getting addicts to become completely drug-free remained the goal for some in the treatment community, numbers of methadone prescriptions increased, and addiction treatment resources shifted from the more abstinence-oriented clinics toward community agencies that were inclined to use harm-reduction measures. These developments were a step away from the British addiction treatment trends of the 1970s and 1980s, which were drifting toward briefer, abstinence-oriented maintenance treatments. Even though British drug policy certainly made a shift toward harm reduction because of AIDS, the changes it brought about were not completely novel. Rather, they marked a reemergence of trends in British drug policy—particularly those of trying to limit the damages caused by drug use and addiction—that had coexisted with control efforts since the 1920s.[8]

In France, the AIDS crisis brought about a more dramatic about-face in addiction treatment policy. Due, in part, to the lack of harm-reduction services—as well as the 1970 law's provisions that made it difficult for treatment providers to advise drug users on how to use safely—HIV infection rates were relatively high among injection-drug users in France. In spite of evidence pointing to the public health dangers inherent in such strict drug policy and treatment approaches, the majority of French treatment providers and policymakers remained strongly opposed to harm-reduction measures, viewing them as a retreat in the campaign against drug use and addiction. Gradually, however, resistance began to thaw in the late 1980s. By the early 1990s, several organizations and a group of general practitioners began to publicly advocate for the expansion of harm-reduction programs. By 1993, France started to expand the availability of substitution treatment and needle exchange, and government attitudes concerning harm reduction began to change in the middle of the decade.[9]

The decisive switch came in 1995, when a government commission headed by medical professor Roger Henrion explored the addiction problem in France. The commission concluded that there was a direct correlation between resistance to harm reduction and high rates of HIV infection among injection-drug users. Not mincing words in its assessment, the Henrion Commission concluded that France's approach to addiction treatment had "led to a health and social catastrophe," and called for dramatic changes. The resulting transformation of the addiction treatment system came swiftly and dramatically, marking a new "French revolution" in the world of addiction treatment. In 1993, only fifty-two people in the entire nation received substitution treatment with methadone or buprenorphine; by 2002, that number had skyrocketed to nearly ninety thousand, the largest in Europe. The swing toward harm reduction changed practices, not just in the addiction treatment community, but in law enforcement. While overall drug-related arrest rates continued to rise, the number of arrests for heroin offenses began to drop in the early 1990s. In 1999, the Minister of Justice issued guidelines encouraging police to issue warnings or to direct addicts to health services instead of arresting them for offenses related to use. These developments reversed France's nearly century-long trend of tackling addiction with tough law-and-order measures while neglecting to address the medical side of the drug problem.[10]

As the nature of drug use and its consequences shifted in the late twentieth century, British and French drug policies responded. In Britain, when it
became clear that the regulations on maintenance prescribing were leading
to a spread of addiction in the 1960s, the government tightened the rules
for treating addiction, and treatment policies and practices became stricter.
When the AIDS crisis made it clear that strict policies could have dire public health consequences, drug policy aims moved back toward reducing the
harms associated with opiate use rather than eliminating it altogether—just
as in the first half of the twentieth century. In France, drug policy followed
a similar trajectory of increasing stringency in the early 1970s, then a shift
toward harm reduction in response to AIDS. But unlike in Britain, where the
changes brought about by AIDS were more subtle and were in line with drug
policy traditions, the French response to AIDS was unprecedented, in terms
of both its scale and its abrupt divergence from the past. Thus, whereas in
Britain there were some continuities between the drug policies of the early
twentieth and early twenty-first centuries, in France there was been a dramatic break, with AIDS facilitating not just a simple shift in drug policy but
a complete upheaval.

Nevertheless, vestiges of what made British and French drug policies distinctive before the 1960s remain. Though it is exceedingly rare, some addicts
still receive maintenance treatment with heroin in Britain today, just as in the
1920s, and heroin is still used in the treatment of severe and intractable pain
even for non-addicts. In France, though the treatment system has expanded
dramatically, the tough measures of the 1970 law regarding "incitement" remain on the books. This has led to some bizarre incidents, such as a 1998
police raid on the Body Shop in Aix because the store had a hemp leaf displayed in its front window, and the 2003 trial of an individual who "incited"
drug use by posting on his website some tips on how to ingest drugs safely.
French politicians have also continued to invoke traditional metaphors of
security and military struggle in speeches about illicit drugs, following the
lead not only of their hard-line "drug warrior" contemporaries in the United
States but also of their French predecessors from a century ago. Just as their
forefathers pointed an accusing finger at other countries (mainly Germany)
when discussing their nation's drug problem in the World War I era, some
French politicians have continued to castigate foreign powers—particularly
the Dutch and their liberal drug policies—in more recent discussions about
drug trafficking in France. Although the drug problem, the overall health
and social consequences of drug use, and policies on addiction have changed,

some remnants of early-twentieth-century drug policies and discourses remain intact in Britain and France as the twenty-first century begins.[11]

Addiction and Citizenship: An Emerging Paradigm?

The growth of addiction as a social issue and the emergence of the AIDS crisis have changed the stakes of addressing opiate addiction and its consequences. Yet, some late-nineteenth- and early-twentieth-century ideas on the connection between drugs and citizenship continue to permeate the way in which addiction is viewed and treated today. This is especially the case in the United States, where treatment providers have integrated "citizens' interventions" into services for individuals recovering from substance use disorders. Recently, leaders in the addiction treatment community have started to follow suit. In 2007, the Betty Ford Institute convened an expert panel to consider how to define recovery from substance use disorders. Beyond sobriety and personal health, the panel added a third dimension—citizenship—to its equation for recovery and the lifestyle that individuals overcoming addiction should try to lead.[12]

Defining citizenship as "working towards the betterment of one's community . . . participating in the rights and responsibilities of social life," and "living with regard and respect for those around you," the Betty Ford panel articulated an understanding of addiction similar to those discussed throughout this book: as a disorder that compromises individuals' capacities to fulfill their obligations to their fellow citizens and the social whole. William L. White, a member of the panel, has elaborated on the decision to include citizenship as an element of recovery, explaining that "there's nothing we know that so transforms personal character and violates relationships to the community [as addiction]." The disorder, he explains, "inevitably involves a process of self-encapsulation" and "drawing into oneself" that often leads to "the abandonment of one's connection and commitment to the community." Consequently, as he told an interviewer from *Time Healthland*, addicts tend to inflict "wounds . . . on the community," and the process of recovery requires a rebuilding of the bonds between drug users and the people around them.[13]

Thus, it would seem, drugs harm addicts physically and psychologically, but also socially and politically, by warping their relationships with their cohort and with society as a whole. Understood to be poisonous to body and mind, habit-forming drugs are also still seen, to a great extent, as socially poisonous as well.

Particular national understandings of citizenship and how it relates to re-covery from addiction are becoming manifest as the Betty Ford Institute's definition is being adapted outside the United States. In 2008, the U.K. Drug Policy Commission, using the Betty Ford Institute's definition as a guide, made reference to the ideals of citizenship in its vision of what recovery from addiction should mean, defining "re-entry into society" and rediscovery of a "productive and meaningful role" within it as essential to overcoming substance use disorders. In particular, the commission found that "work" should constitute a key element in recovery for many addicts—a conclusion that echoes British conceptions of citizenship and its relationship to drug use from the early twentieth century.[14]

According to White and his colleagues, "we are just beginning to understand this third dimension, citizenship, and how it can best be enhanced and measured" as a component of recovery from substance use disorders. "This frontier," they conclude, "is a worthy arena for research and service experimentation."[15] What role should ideals of citizenship play in the development of addiction treatment services and policies in the twenty-first century? Treatment providers across the world are just now beginning to consider this question in detail. However, as I have tried to show in this book, ideals of citizenship colored the way industrialized societies addressed the drug problem from its beginnings. Indications are that they will continue to do so in the future as well.

Introduction • A Tale of Two Drug Policies

1. Jan-Willem Gerritsen, *The Control of Fuddle and Flash: A Sociological History of the Regulation of Alcohol and Opiates* (Leiden, Netherlands: Brill, 2000).

2. Gerald M. Oppenheimer, "To Build a Bridge: The Use of Foreign Models by Domestic Critics of U.S. Drug Policy," *Milbank Quarterly* 69, no. 3 (1991), 495–512; John Strang and Michael Gossop, "The 'British System': Visionary Anticipation or Masterly Inactivity?" in John Strang and Michael Gossop (eds.), *Heroin Addiction and Drug Policy: The British System* (Oxford: Oxford University Press, 1994), 342; Alex Mold, *Heroin: The Treatment of Addiction in Twentieth-Century Britain* (DeKalb, IL: Northern Illinois University Press, 2008), 9–10. For examples of critics using Britain as a comparative foil in discussions of other nations' policies, see Alfred Lindesmith, *The Addict and the Law* (Bloomington: Indiana University Press, 1965), 162–188; Edwin M. Schur, *Narcotic Addiction in Britain and America: The Impact of Public Policy* (Bloomington: Indiana University Press, 1968); Arnold S. Trebach, *The Heroin Solution* (New Haven, CT: Yale University Press, 1982), 85–117.

3. In addition to the works mentioned in the previous note, see Christian Bachmann and Anne Coppel, *Le dragon domestique: deux siècles de relations étranges entre l'Occident et la drogue* (Paris: Albin Michel, 1989), 287.

4. Virginia Berridge, "The 'British System' and Its History: Myth and Reality," in John Strang and Michael Gossop (eds.), *Heroin Addiction and the British System, Volume One: Origins and Evolution* (London: Routledge, 2005), 15–16; Bing Spear, "The Early Years of the 'British System' in Practice," in Strang and Gossop, *Heroin Addiction and Drug Policy*, 3–28; Terry M. Parssinen, *Secret Passions, Secret Remedies: Narcotic Drugs in British Society, 1820–1930* (Philadelphia: Institute for the Study of Human Issues, 1983), 212–220; Emmanuelle Retaillaud-Bajac, *Les paradis perdus: drogues et usagers de drogues dans la France de l'entre-deux-guerres* (Rennes, France: Presses Universitaires de Rennes, 2009), 414, 416; John Witton, Francis Keaney, and John Strang, "They Do Things Differently over There: Doctors, Drugs, and the 'British System' of Treating Opiate Addiction," *Journal of Drug Issues* 35, no. 4 (2005), 781; Mold, *Heroin*, 10.

5. Throughout the book, I follow the nomenclature used by other historians, using the term *opiates* as shorthand to describe opium itself, morphine (an alkaloid found in opium), heroin (a semisynthetic drug made from morphine), and other opiates and semisynthetic opiates. Darryl

S. Inaba and William E. Cohen, *Uppers, Downers, All Arounders: Physical and Mental Effects of Psychoactive Drugs* (Medford, OR: CNS Publications), 141–142.

6. Bachmann and Coppel, *Le dragon domestique*, 287; Retaillaud-Bajac, *Les paradis perdus*, 414, 416.

7. Parssinen, *Secret Passions, Secret Remedies*, 68–78; Gerritsen, *Control of Fuddle and Flash*, 143; Richard Davenport-Hines, *The Pursuit of Oblivion: A Global History of Narcotics, 1500–2000* (London: Weidenfeld & Nicolson, 2001), 91–92, 155–156; David F. Musto, *The American Disease: Origins of Narcotics Control* (New York: Oxford University Press, 1987), 5–6; Louise Foxcroft, *The Making of Addiction: The "Use and Abuse" of Opium in Nineteenth-Century Britain* (Aldershot, UK: Ashgate, 2007), 79–169; Virginia Berridge, *Opium and the People: Opiate Use and Drug Control Policy in Nineteenth and Early Twentieth Century England* (London: Free Association Books, 1999), 135–170; Jean-Jacques Yvorel, *Les poisons de l'esprit: drogues et drogués au XIXᵉ siècle* (Paris: Quai Voltaire, 1992), 49–68, 78–93; David T. Courtwright, *Dark Paradise: Opiate Addiction in America before 1940* (Cambridge, MA: Harvard University Press), 81–83; David T. Courtwright, *Forces of Habit: Drugs and the Making of the Modern World* (Cambridge, MA: Harvard University Press, 2001), 168–169.

8. Rachel Lart, "Medical Power and Knowledge: The Treatment and Control of Drugs and Drug Users," in Ross Coomber (ed.), *The Control of Drugs and Drug Users: Reason or Reaction?* (Amsterdam: Harwood Academic Publishers, 1998), 49–56; Geoffrey Harding, *Opiate Addiction, Morality, and Medicine: From Moral Illness to Pathological Disease* (New York: St. Martin's Press, 1988); Marcus Aurin, "Chasing the Dragon: The Cultural Metamorphosis of Opium in the United States, 1825–1935," *Medical Anthropology Quarterly*, new ser., 14, no. 3 (2000), 415. Conrad and Schneider argue that opiate addiction was "medicalized" and then subsequently "demedicalized" with the advent of tighter controls over narcotics in the early twentieth century. Peter Conrad and Joseph W. Schneider, *Deviance and Medicalization: From Badness to Sickness* (Philadelphia: Temple University Press, 1992), 110–144.

9. Courtwright, *Dark Paradise*, 78–79, 113–147; Courtwright, *Forces of Habit*, 171; Parssinen, *Secret Passions, Secret Remedies*, 115–128; Musto, *American Disease*, 5–6, 11, 20, 28–40, 43, 49–53; Foxcroft, *Making of Addiction*, 63–65, 71–72; Catherine Carstairs, *Jailed for Possession: Illegal Drug Use, Regulation, and Power in Canada, 1920–1961* (Toronto: University of Toronto Press, 2006), 17–34; Timothy A. Hickman, *The Secret Leprosy of Modern Days: Narcotic Addiction and Cultural Crisis in the United States, 1870–1920* (Amherst: University of Massachusetts Press, 2007), 60–92; William B. McAllister, *Drug Diplomacy in the Twentieth Century: An International History* (London: Routledge, 2000), 17, 20–21, 27–35; David Bewley-Taylor, *The United States and International Drug Control, 1909–1997* (London: Pinter, 1999), 17–18; Davenport-Hines, *Pursuit of Oblivion*, 87–90; Berridge, *Opium and the People*, 97–109, 195–205.

10. Berridge, *Opium and the People*, 113–131; Caroline Jean Acker, "From All Purpose Anodyne to Marker of Deviance: Physicians' Attitudes towards Opiates in the US from 1890–1940," in Roy Porter and Mikulas Tiech (eds.), *Drugs and Narcotics in History* (Cambridge: Cambridge University Press, 1995), 114–132; Parssinen, *Secret Passions, Secret Remedies*, 69–72; Musto, *American Disease*, 13–16, 56–58; Hickman, *Secret Leprosy of Modern Days*, 93–124; Timothy Hickman, "The Double Meaning of Addiction: Habitual Narcotic Use and the Logic of Professionalizing Medical Authority in the United States, 1900–1920," in Sarah W. Tracy and Caroline Jean Acker (eds.), *Altering American Consciousness: The History of Alcohol and Drug Use in the United States,*

1800–2000 (Amherst: University of Massachusetts Press, 2004), 182–202; Aurin, "Chasing the Dragon," 415–417, 435.

11. Musto, *American Disease*, 134; Caroline Jean Acker, *Creating the American Junkie: Addiction Research in the Classic Era of Narcotic Control* (Baltimore: Johns Hopkins University Press, 2002), 33; Robert P. Stephens, *Germans on Drugs: The Complications of Modernization in Hamburg* (Ann Arbor: University of Michigan Press, 2007), 16–17; Courtwright, *Forces of Habit*, 171–173; Yvorel, *Les poisons de l'esprit*, 217–220; Zhou Yongming, *Anti-Drug Crusades in Twentieth Century China: Nationalism, History, and State Building* (Lanham, MD: Rowman & Littlefield Publishers, 1999); Susan L. Speaker, "The Struggle of Mankind against Its Deadliest Foe: Themes of Counter-Subversion in Anti-Narcotic Campaigns," *Journal of Social History* 34, no. 3 (2001), 591–610.

12. Peter Reuter, Mathea Falco, and Robert MacCoun, *Comparing Western European and North American Drug Policies: An International Conference Report* (Santa Monica, CA: RAND Corporation, 1993), 24; Harald Klingemann and Geoffrey Hunt (eds.), *Drug Treatment Systems in an International Perspective: Drugs, Demons, and Delinquents* (Thousand Oaks, CA: Sage Publications, 1998); Thomas Babor, Jonathan Caulkins, Griffith Edwards, Benedikt Fischer, David Foxcroft, Keith Humphreys, Isidore Obot, Jürgen Rehm, Peter Reuter, Robin Room, Ingebort Rossow, and John Strang, *Drug Policy and the Public Good* (Oxford: Oxford University Press, 2010), 221; Tom Decorte and Dirk Korf (eds.), *European Studies on Drugs and Drug Policy* (Brussels: VUB Press, 2004); Nicholas Dorn, Jorgen Jepsen, and Ernesto Savona (eds.), *European Drug Policies and Enforcement* (Houndmills, UK: Macmillan Press, 1996); Ulrik Solberg, Gregor Burkhart, and Margareta Nilson, "Opiate Substitution Treatment in the European Union and Norway," *International Journal of Drug Policy* 13, no. 6 (2002), 477–484; Peter Baldwin, *Disease and Democracy: The Industrialized World Faces AIDS* (Berkeley: University of California Press, 2005), 147–151.

13. Baldwin, *Disease and Democracy*, 152; Ellen Benoit, "Not Just a Matter of Criminal Justice: States, Institutions, and North American Drug Policy," *Sociological Forum* 18, no. 2 (2003), 269–294; Robert J. MacCoun and Peter Reuter, *Drug War Heresies: Learning from Other Vices, Times & Places* (Cambridge: Cambridge University Press, 2001), 209–210; Babor et al., *Drug Policy and the Public Good*, 227–233; Josef Radimecký, "Rhetoric versus Practice in Czech Drug Policy," *Journal of Drug Issues* 37, no. 1 (2007), 11–44; Tim Boekhout van Solinge, *Dealing with Drugs in Europe: An Investigation of European Drug Control Experiences—France, the Netherlands, and Sweden* (The Hague: Boom Juridische uitgevers, 2004), 29, 213–214.

14. Boekhout van Solinge, *Dealing with Drugs in Europe*, 1–6, 189–192; Alain Ehrenberg, *L'individu incertain* (Paris: Calmann-Levy, 1995), 70–71; Alain Ehrenberg, "Comment vivre avec les drogues? Questions de recherche et enjeux politiques," *Communications* 62 (1996), 8–9; Peter Baldwin, "Beyond Weak and Strong: Rethinking the State in Comparative Policy History," *Journal of Policy History* 17, no. 1 (2005), 28; Babor et al., *Drug Policy and the Public Good*, 232–233; Marcel de Kort and Ton Cramer, "Pragmatism versus Ideology: Dutch Drug Policy Continued," *Journal of Drug Issues* 29, no. 3 (1999), 476–477; Ted Goldberg, "The Evolution of Swedish Drug Policy," *Journal of Drug Issues* 34, no. 3 (2004), 551–576; Tuukka Tammi, "Discipline or Contain? The Struggle over the Concept of Harm Reduction in the 1997 Drug Policy Committee in Finland," *International Journal of Drug Policy* 16, no. 6 (2005), 384–392.

15. Boekhout van Solinge, *Dealing with Drugs in Europe*, 29.

16. Ibid., 5–6, 210–211, 218.

17. "L'opium," *Le Matin*, June 2, 1913.

18. Thomas De Quincey, "Confessions of an English Opium Eater: Being an Extract from the Life of a Scholar" (1821), in Thomas De Quincey (ed. Barry Milligan), *Confessions of an English Opium Eater and Other Writings* (London: Penguin Books, 2003); Departmental Committee on Morphine and Heroin Addiction, *Report* (London: His Majesty's Stationery Office, 1926). For discussions of mid-nineteenth-century British and French pharmaceutical regulations, see Yvorel, *Les poisons de l'esprit*, 56–57; and Berridge, *Opium and the People*, xv, 75–121.

19. Royal Commission on Opium, *Final Report* (London: Her Majesty's Stationery Office, 1895); "L'affaire Ullmo: un procès de haute trahison," *Revue des Grands Procès Contemporains: Recueil d'Éloquence Judiciaire* 36 (1908), 209–260.

20. Departmental Committee on Morphine and Heroin Addiction, *Report*.

21. See Jal Mehta, "The Varied Roles of Ideas in Politics: From 'Whether' to 'How,'" in Daniel Béland and Robert Henry Cox (eds.), *Ideas and Politics in Social Science Research* (Oxford: Oxford University Press, 2011), 27–38; Vivien A. Schmidt, "Discursive Institutionalism: The Explanatory Power of Ideas and Discourse," *Annual Review of Political Science* 11 (2008), 306, 308; Frank Dobbin, *Forging Industrial Policy: The United States, Britain, and France in the Railway Age* (Cambridge: Cambridge University Press, 1994), 20–23; Adrian Favell, *Philosophies of Integration: Immigration and the Idea of Citizenship in France and Britain* (New York: St. Martin's Press, 1998), 16; Mark Blyth, *Great Transformations: Economic Ideas and Institutional Change in the Twentieth Century* (Cambridge: Cambridge University Press, 2002), 267–271; Boekhout van Solinge, *Dealing with Drugs in Europe*, 211–213.

22. Boekhout van Solinge, *Dealing with Drugs in Europe*, 211–213.

Chapter One • Imagining the Meditative Nation

Epigraph. Charles Baudelaire (trans. Stacy Diamond), *Artificial Paradises* (1860) (New York: Carol Publishing Group, 1996), 24.

1. Roy Porter, "Introduction," in Jean-Charles Sournia (trans. Nick Hindley and Gareth Stanton), *A History of Alcoholism* (Oxford: Basil Blackwell, 1990), x; Jean-Charles Sournia, *A History of Alcoholism* (Oxford: Basil Blackwell, 1990), 6; Gregory A. Austin, *Alcohol in Western Society from Antiquity to 1800* (Santa Barbara CA: ABC-CLIO Information Services, 1985), xvi–xvii, 274–275; Louise Hill Curth, "The Medicinal Value of Wine in Early Modern England," *Social History of Alcohol and Drugs* 18 (2003), 35–50; Mikhail Bakhtin (trans. Helene Iswolsky), *Rabelais and His World* (Bloomington: Indiana University Press, 1984), 75; Joni B. McNutt, *In Praise of Wine: An Offering of Hearty Toasts, Quotations, Witticisms, Proverbs, and Poetry throughout History* (Santa Barbara, CA: Capa Press, 1993), 21, 40, 45.

2. On the social rites surrounding alcohol consumption and the cultural significance of drinking, see Wolfgang Schivelbusch (trans. David Jacobson), *Tastes of Paradise: A Social History of Spices, Stimulants, and Intoxicants* (New York: Pantheon Books, 1992), 78, 169–171; Marianna Adler, "From Symbolic Exchange to Commodity Consumption: Anthropological Notes on Drinking as a Symbolic Practice," in Susanna Barrows and Robin Room (eds.), *Drinking: Behavior and Belief in Modern History* (Berkeley: University of California Press, 1991), 381–384; Thomas Brennan, "Social Drinking in Old Regime Paris," in Barrows and Room, *Drinking*, 75;

Didier Nourrisson, *Le buveur du XIXe siècle* (Paris: Albin Michel, 1990), 115. On the evolution of drinking spaces as centers of sociability, see Susanna Barrows and Robin Room, "Introduction," in Barrows and Room, *Drinking*, 8; Austin, *Alcohol in Western Society*, 277; Schivelbusch, *Tastes of Paradise*, 149, 188, 194; W. Scott Haine, "From Drinkseller to Social Entrepreneur: The Parisian Working-Class Café Owner, 1789–1914," in Jack S. Blocker and Cheryl Krasnick Warsh (eds.), *The Changing Face of Food and Drink: Substance, Imagery, and Behavior* (Ottawa: Social History, 1997), 99–104.

3. Baudelaire, *Artificial Paradises*, 4; Defoe quoted in Austin, *Alcohol in Western* Society, 274–275; McNutt, *In Praise of Wine*, 198; Griffith Edwards, *Alcohol: The Ambiguous Molecule* (London: Penguin Books, 2000), 18–22; Adler, "From Symbolic Exchange to Commodity Consumption," 381.

4. Edwards, *Alcohol*, 47; Thomas Nashe, *Pierce Penilesse: His Svpplication to the Divell* (1592) (London: John Lane the Bodley Head, 1924), 77–80; *A Treatise on the Effects of Drinking Spirituous Liquors, Wine and Beer, on the Body and Mind* (London, 1794), 11; Stephen Hales, *A Friendly Admonition to the Drinkers of Gin, Brandy, and Other Distilled Spirituous Liquors* (London: B. Dod, 1754), 13; Thomas Trotter, *An Essay, Medical, Philosophical, and Chemical, on Drunkenness and Its Effects on the Human Body*, 4th ed. (London: Longman, Hurst, Rees and Orme, 1810), 20.

5. Trotter, *Essay, Medical, Philosophical*, 42, 59–62; Robert Nye, *Crime, Madness & Politics in Modern France: The Medical Concept of National Decline* (Princeton, NJ: Princeton University Press, 1985), 124, 156; Patricia E. Prestwich, *Drink and the Politics of Social Reform: Antialcholism in France since 1870* (Palo Alto, CA: Society for the Promotion of Science and Scholarship, 1988), 39; "The Action of Alcohol," *British Medical Journal*, April 4, 1874, 457–458; Lionel S. Beale, "On 'Deficiency of Vital Power' in Disease, and on 'Support': With Observations upon the Action of Alcohol in Serious Cases of Acute Disease," *British Medical Journal*, October 10, 1863, 389–390; Robert Jones, "The Relation of Inebriety to Mental Diseases," *British Journal of Inebriety* 2, no. 1 (1904), 5–6, 9; T. D. Crothers, "A Contribution to the Pathology of Inebriety," *British Journal of Inebriety* 1, no. 2 (1903), 82–87; A. T. Shearman, "The Mental Development of the Inebriate," *British Journal of Inebriety* 1, no. 4 (1904) 259; Norman Kerr, *Inebriety: Its Etiology, Pathology, Treatment, and Jurisprudence*, 2nd ed. (London: H. K. Lewis, 1889), 65–67.

6. Prestwich, *Drink and the Politics of Social Reform*, 14, 43–51; John Greenaway, *Drink and British Politics since 1830: A Study in Policy-Making* (Houndmills, UK: Palgrave Macmillan, 2003), 8; Austin, *Alcohol in Western Society*, 7–8, 20.

7. Kerr, *Inebriety*, 49–52.

8. A. T. Shearman, "The Effect of Alcohol on Feeling," *British Journal of Inebriety* 3, no. 4 (1906), 185–186; Kerr, *Inebriety*, 96; John F. Nicolson, "The Use of Alcohol," *British Medical Journal*, December 7, 1861, 621–622; Edwin Lankester, "The Action of Alcohol," *British Medical Journal*, November 9, 1861, 515.

9. Mordecai C. Cooke, *The Seven Sisters of Sleep: Popular History of the Seven Prevailing Narcotics of the World* (Lincoln, MA: Quarterman Publications, 1989), 118–121; Mike Jay, *Emperors of Dreams: Drugs in the Nineteenth Century* (Sawtry, UK: Dedalus 2000), 66; Virginia Berridge, *Opium and the People: Opiate Use and Drug Control Policy in Nineteenth and Early Twentieth Century England* (New York: Free Association Books, 1999), 3; Guibourt, "Opium, histoire et description," in Gabriel Andral, Louis Jacques Bégin, and Philippe Frédéric Blandin,

Dictionnaire de médecine et de chirurgie pratique (Paris: J. B. Baillière, 1834), 12: 213; James Rennie, *A New Supplement to the Latest Pharmacopoeias of London, Edinburgh, Dublin, and Paris* (London: Baldwin and Cradock, 1837), 291; Jean-Baptiste Fonssagrives, "Opium," in A. Dechambre, *Dictionnaire encyclopédique des sciences médicales*, 2nd ser. (Paris: Imprimerie A. Lahure, 1881), 16: 151–152; Martin Booth, *Opium: A History* (New York: St. Martin's Press, 1996), 19–26.

10. Booth, *Opium*, 26–27, 58–59; David Musto, *The American Disease: Origins of Narcotic Control* (New York: Oxford University Press, 1987), 69–70; Fonssagrives, "Opium," 160; David Courtwright, *Dark Paradise: Opiate Addiction in America before 1940* (Cambridge, MA: Harvard University Press, 1982), 43; Jean-Jacques Yvorel, *Les poisons de l'esprit: drogues et drogués au XIXe* *siècle* (Paris: Quai Voltaire, 1992), 50–53; Berridge, *Opium and the People*, xx, 24–28; Terry M. Parssinen, *Secret Passions, Secret Remedies: Narcotic Drugs in British Society, 1820–1930* (Philadelphia: Institute for the Study of Human Issues, 1983), 30–36; R. Hooper, *A Compendious Medical Dictionary* (Boston: Manning and Loring, 1801), 200.

11. Alethea Hayter, *Opium and the Romantic Imagination* (London: Faber and Faber, 1968), 25, 29; George Young, *A Treatise on Opium, Founded upon Practical Observations* (London: A. Millar, 1753), 14, 169; Rennie, *New Supplement*, 292; Samuel Crumpe, *An Inquiry into the Nature and Properties of Opium* (London: G. G. and J. Robinson, 1793), 31, 37, 89, 93–97.

12. Crumpe, *Inquiry into Nature and Properties*, 48; Martin Solon, "Opium (thérapeutique)," in Andral et al., *Dictionnaire de médecine et de chirurgie pratique*, 235; Young, *Treatise on Opium*, 108; Hayter, *Opium and the Romantic Imagination*, 28; P. A. Charvet, *De l'action comparée de l'opium et de ses principes continuant sur l'économie animale* (Paris: G. G. Levrault, 1826), 60; Baron de Tott, *Mémoires du Baron de Tott sur les Turcs et les Tartares* (Paris, 1785), 1: 88. Captain Cook's description of Malay opium users is discussed in Marcus Boon, *The Road of Excess: A History of Writers on Drugs* (Cambridge, MA: Harvard University Press, 2002), 22.

13. Thomas Wharton Jr. quoted in Boon, *Road of Excess*, 19–20; Crumpe, *Inquiry into Nature and Properties*, 45–47; Berridge, *Opium and the People*, 38–48, 97–117; Parssinen, *Secret Passions, Secret Remedies*, 46; Yvorel, *Les poisons de l'esprit*, 55–56.

14. Max Milner, *L'imaginaire des drogues: de Thomas de Quincey à Henri Michaux* (Paris: Éditions Gallimard, 2000), 8; Patrick Walberg, "Foreword," in Arnould de Liedekerke, *La belle époque de l'opium* (Paris: Éditions de Différence, 2001), 11–12; Yvorel, *Les poisons de l'esprit*, 31–39; Berridge, *Opium and the People*, 214–215; Boon, *Road of Excess*, 6, 30–31. Coleridge's correspondence with Wedgwood is discussed in Jay, *Emperors of Dreams*, 57–62, 105–113, 140–143.

15. Elisabeth Schneider, *Coleridge, Opium, and Kubla Khan* (Chicago: University of Chicago Press, 1953), 21–26; Boon, *Road of Excess*, 35.

16. Thomas De Quincey, "Confessions of an English Opium Eater: Being an Extract from the Life of a Scholar" (1821), in Thomas De Quincey (ed. Barry Milligan), *Confessions of an English Opium Eater and Other Writings* (London: Penguin Books, 2003), 47; Thomas De Quincey, "Suspiria de Profundis: Being a Sequel to the Confessions of an English Opium Eater" (1845), in De Quincey, *Confessions of an English Opium Eater*; Thomas De Quincey, "The English Mail Coach" (1849), in De Quincey, *Confessions of an English Opium Eater*.

17. Barry Milligan, "Introduction," in De Quincey, *Confessions of an English Opium Eater*, xviii; De Quincey, "Suspiria de Profundis," 150, 152.

18. De Quincey, "Confessions of an English Opium Eater," 77, 81–82; Timothy A. Hickman, *The Secret Leprosy of Modern Days: Narcotic Addiction and Cultural Crisis in the United States,*

1870–1920 (Amherst: University of Massachusetts Press, 2007), 17–19; Barry Milligan, *Pleasures and Pains: Opium and the Orient in Nineteenth-Century British Culture* (Charlottesville: University of Virginia Press, 1995), 12, 48–49.

19. De Quincey, "Confessions of an English Opium Eater," 55, 76; De Quincey, "Suspiria de Profundis," 110; Hickman, *Secret Leprosy of Modern Days*, 19.

20. De Quincey, "Suspiria de Profundis," 89–91, emphasis in original.

21. De Quincey, "Confessions of an English Opium Eater," 7–10, 24, 47–50, 270–271, emphases in original.

22. Ibid., 45–47, emphases in original.

23. Boon, *Road of Excess*, 41; Berridge, *Opium and the People*, 15, 55; Parssinen, *Secret Passions, Secret Remedies*, 5; Milligan, *Pleasures and Pains*, 8; Milner, *L'imaginaire des drogues*, 58; Liedekerke, *La belle époque de l'opium*, 11, 62; Hickman, *Secret Leprosy of Modern Days*, 1, 15–22, 25–32; George Vigarello, "La drogue a-t-elle un passé?" in Alain Ehrenberg (ed.), *Individus sous influence: drogues, alcools, médicaments, psychotropes* (Paris: Éditions Esprit, 1991), 93; Hayter, *Opium and the Romantic Imagination*, 233; Laurent Tailhade, *Omar Khayyam et les poisons de l'intelligence* (Paris: Charles Carrington, 1905), 75; Baudelaire, *Artificial Paradises*. For examples of case studies in the 1920s and 1930s that mention De Quincey and addicts citing him as their inspiration to use opiates, see G. Laughton Scott, *The Morphine Habit and Its Painless Treatment* (London: H. K. Lewis, 1930), 10; Jean Goudot, *Quelques considérations sur le traitement et le pronostic des toxicomanies* (Paris: Librairie E. le François, 1936), 1.

24. Throughout the rest of this chapter, my analysis is based on the written evidence left by doctors, artists, and men of letters. The boundaries between the science and art of opium were often fluid in the nineteenth century, with scientific writing informing cultural perceptions of opium, and vice versa. Yvorel, *Les poisons de l'esprit*. 44; Virginia Berridge, "The Origins of the English Drug 'Scene' 1890–1930," *Medical History* 32, no. 1 (1988), 58.

25. "Noctes Ambrosianae," *Lancet*, August 7, 1830, 749; Henri Georgelin, *Étude sur l'opiomanie et les fumeurs d'opium: considérés au point de vue de l'hygiène sociale* (Bordeaux: Imprimerie Commerciale et Industrielle, 1906), 30; Fonssagrives, "Opium," 164, 183, 189; Richard Millant, *La drogue: fumeurs et mangeurs d'opium* (Paris: Librairie Africaine & Coloniale, 1910), 34; George G. Sigmond, "Lectures on Materia Medica and Therapeutics, Lecture VI," *Lancet*, December 3, 1836, 357; Hayter, *Opium and the Romantic Imagination*, 46; Cooke, *Seven Sisters of Sleep*, 151.

26. Hayter, *Opium and the Romantic Imagination*, 45–46; *Lancet*, "Noctes Ambrosianae," 750; Sigmond, "Lectures on Materia Medica, VI," 357; Georgelin, *Étude sur l'opiomanie*, 30; Roger Dupouy, *Les opiomanes: mangeurs, buveurs et fumeurs d'opium* (Paris: Librairie Félix Alcan, 1911), 24, 117.

27. Fonssagrives, "L'opium," 182; P. Brouardel, *Opium, morphine, et cocaïne: intoxication aigue par l'opium, mangeurs et fumeurs d'opium, morphinomanes et cocaïnomanes* (Paris: J. B. Baillière, 1906), 35, 47; Pierre Loti, *Les derniers jours de Pékin* (1901) (Paris: Calmann-Lévy, 1925), 237; John Simon, "A Course of Lectures in General Pathology; Delivered at St. Thomas's Hospital, Lecture X," *Lancet*, August 10, 1850, 170; S.S., "The Effects of Small Doses of Opium," *Lancet*, March 21, 1863, 341; Cooke, *Seven Sisters of Sleep*, 133, 137; "Opium Eating," *Lancet*, August 22, 1835, 685; Claude Farrère (trans. Samuel Putnam), *Black Opium* (1906) (New York: Nicholas L. Brown, 1929), 172; Robert Little, "On the Habitual Use of Opium," *Monthly Journal of Medical*

Science 10, no. 6 (June 1850), 527; Charles Richet quoted in Dupouy, *Les opiomanes*, 24; B. A. [Bénédict] Morel, *Traité des dégénérescences physiques, intellectuelles et morales de l'espèce humaine et des causes qui produisent ces variétés maladives* (Paris: J. B. Baillière, 1857), 163.

28. Georgelin, *Étude sur l'opiomanie*, 35–36; Morel, *Traité des dégénérescences*, 163; *Lancet*, "Noctes Ambrosianae," 748; Millant, *La drogue*, 23; Tailhade, *Omar Khayyam*, 75; Brouardel, *Opium, morphine et cocaïne*, 48; J. Luys, "Impressions d'un buveur d'opium," *L'Encéphale* 7 (1887), 306; Kerr, *Inebriety*, 106; Baudelaire, *Artificial Paradises*, 102; Dupouy, *Les opiomanes*, 20, 27–29, 85–86, 94.

29. Fonssagrives, "L'opium," 182; Cooke, *Seven Sisters of Sleep*, 140; Luys, "Impressions d'un buveur d'opium," 306; Dupouy, *Les opiomanes*, 24, 86–89; Baudelaire, *Artificial Paradises*, 119, emphasis in original; Farrère, *Black Opium*, 49–50, 164, 188.

30. Farrère, *Black Opium*, 36–37, 50; Milner, *L'imaginaire des drogues*, 203; Arthur Symons, "The Opium Smoker," *Academy* 32 (Nov. 19, 1887), 336; Dupouy, *Les opiomanes*, 115; Baudelaire, *Artificial Paradises*, 157; Pierre Pachet, "Coleridge, De Quincey, Baudelaire: la drogue de l'individu moderne," in Ehrenberg, *Individus sous l'influence*, 38; Sigmond, "Lectures on Materia Medica, VI," 356.

31. Dupouy, *Les opiomanes*, 81; A. [Albert de] Pouvourville (pseudonym, Matgioi), *Les éléments de l'homme: la pathogène chinoise* (Paris: Amuel, 1895), 46, 50; *Lancet*, "Opium Eating," 685; Alain Ehrenberg, *L'individu incertain* (Paris: Calmann-Lévy, 1995), 49; "Proceedings of Westminster Medical Society, December 21 1839," *Lancet*, January 4, 1840, 545.

32. Millant, *La drogue*, 24; Hayter, *Opium and the Romantic Imagination*, 334; Milner, *L'imaginaire des drogues*, 215, 218; Dupouy, *Les opiomanes*, 96–97, 123; Tailhade, *Omar Khayyam*, 75; Alonzo Calkins, *Opium and the Opium Appetite: With Notices of Alcoholic Beverages, Cannabis Indica, Tobacco and Coca, and Tea and Coffee in Their Hygienic Aspects and Pathologic Relations* (Philadelphia: J. B. Lippincott, 1871), 60; Delphi Fabrice, *Opium à Paris* (Paris, 1914), 97; Farrère, *Black Opium*, 164–165, 201.

33. Aleister Crowley, *The Diary of a Drug Fiend* (1922) (York Beach, ME: Samuel Weiser, 1981), 69–71; Jean Cocteau (trans. Ernest Boyd), *Opium: The Diary of an Addict* (1930) (London: George Allen & Unwin, 1933), 72.

34. [Albert de] Pouvourville (pseudonym, Matgioi), *L'esprit des races jaunes: l'opium—sa pratique* (Paris: Librarie Paul Ollendorf, 1902), 14, 54; Tailhade, *Omar Khayyam*, 66; Dupouy, *Les opiomanes*, 121.

35. *Lancet*, "Opium Eating," 684; Loti, *Les derniers jours de Pékin*, 236, 241; "The Opium Habit," *Lancet*, April 7, 1888, 692; Kerr, *Inebriety*, 96; Dupouy, *Les opiomanes*, 60–61; Cooke, *Seven Sisters of Sleep*, 135; Parssinen, *Secret Passions, Secret Remedies*, 66–67, emphasis in original; Farrère, *Black Opium*, 163; Hayter, *Opium and the Romantic Imagination*, 66; Berridge, "Origins of the English Drug 'Scene,'" 53.

36. Baudelaire, *Artificial Paradises*, 60; Tailhade, *Omar Khayyam*, 63, 66.

37. Dupouy, *Les opiomanes*, 98; Baudelaire, *Artificial Paradises*, 102; Calkins, *Opium and the Opium Appetite*, 76; Liedekerke, *La belle époque de l'opium*, 159, 182–183; G. Levett-Yeats, "The Land of the Poppy: Among the Opium Vats," *Macmillan's Magazine* 84 (May/Oct. 1901), 273; Georges Thibout, *La question de l'opium dans l'époque contemporaine* (Paris: G. Steinheil, 1912), 7; Cooke, *Seven Sisters of Sleep*, 148.

38. Milligan, *Pleasures and Pains*, 20–21; Boon, *Road of Excess*, 219; Yvorel, *Les poisons de*

l'esprit, 169–175; Milner, *L'imaginaire des drogues,* 231; Berridge, *Opium and the People,* 173–175; Chantal Descours-Gatin, *Quand l'opium finançait la colonisation en Indochine* (Paris: Éditions L'Harmattan, 1992); Dupouy, *Les opiomanes,* 8.

39. "Ennui and the Opium Pipe: A Chinese View," *Temple Bar* 102 (1894), 507; Cooke, *Seven Sisters of Sleep,* 141; A Competition Wallah, "Letters on India: Letter III about Opium and Other Things," *Macmillan's Magazine* 8 (May/Oct. 1863), 275.

40. Farrère, *Black Opium,* 146; Liedekerke, *La belle époque de l'opium,* 179; Charles Dickens, "Lazarus, Lotus-Eating," *All the Year Round,* May 12, 1866, 425; Daniel Borys, *Le royaume de l'oubli: pathologie et psychologie de fumeurs d'opium* (Paris: Louis-Michaud, 1909), 103; "A Night in an Opium Den," *Strand Magazine* 1 (1891), 625; Susan Zieger, *Inventing the Addict: Drugs, Race, and Sexuality in Nineteenth-Century British and American Literature* (Amherst: University of Massachusetts Press, 2008), 19.

41. David T. Courtwright, "Mr. ATOD's Wild Ride: What Do Alcohol, Tobacco, and Other Drugs Have in Common?" *Social History of Alcohol and Drugs* 20 (2005), 107–109; Virginia Berridge, "The Origins and Early Years of the Society, 1884–1899," *British Journal of Addiction* 85, no. 8 (1990), 1000–1001; Hasso Spode, "Transubstantiations of the Mystery: Two Remarks on the Shifts in the Knowledge about Addiction," *Social History of Alcohol and Drugs* 20 (2005), 127; Yvorel, *Les poisons de l'esprit,* 63–68, 78–93; Emmanuelle Retaillaud-Bajac, *Les paradis perdus: drogues et usagers de drogues dans la France de l'entre-deux-guerres* (Rennes, France: Presses Universitaires de Rennes, 2009), 276; Jean Dugarin and Patrice Nominé, "Toxicomanie: historique et classifications," *Histoire, Économie et Société* 7, no. 4 (1988), 584–585; Morel, *Traité des dégénérescences,* 48–49; Kerr, *Inebriety,* 6, 33–35. Concerns about inebriety were often linked with broader discussions of social and racial degeneration. For an overview of degeneration theories, see Daniel Pick, *Faces of Degeneration: A European Disorder, c. 1848–1918* (Cambridge: Cambridge University Press, 1989).

42. Magnus Huss, *Alcoholismus chronicus, eller chronisk alkoholssjukdom* (Stockholm: Beckman, 1849). For an overview of late-nineteenth- and early-twentieth-century research on alcoholism, see Prestwich, *Drink and the Politics of Social Reform,* 38; Nye, *Crime, Madness & Politics in Modern France,* 156; Berridge, "Origins and Early Years," 999; Virginia Berridge, "Prevention and Social Hygiene, 1900–1914," *British Journal of Addiction* 85, no. 8 (1990), 1009–1013. For an example of early-nineteenth-century thought on the effects of excessive drinking, see Trotter, *Essay, Medical, Philosophical,* 20–26, 45–62.

For examples of late-nineteenth- and early-twentieth-century writing about alcohol in Britain, see Kerr, *Inebriety,* 19–47; Crothers, "Contribution to the Pathology of Inebriety," 81–87; Norman Kerr, "Alcoholism and Drug Habits," in Thomas L. Stedman (ed.), *Twentieth Century Practice: An International Encyclopedia of Modern Medical Science by Leading Authorities in Europe and America* (London: Sampson Low, Martson and Company, 1895), 3: 6–50; Patrick Hehir, *Opium: Its Physical, Moral, and Social Effects* (London: Baillière, Tindall and Co., 1895), 238–239, 242, 246, 253–254; Harry Campbell, "The Study of Inebriety: A Retrospect and Forecast," *British Journal of Inebriety* 1, no. 1 (1903), 7; W. Ford Robertson, "The Pathology of Chronic Alcoholism," *British Journal of Inebriety* 1, no. 4 (1904), 229–235; T. D. Crothers, "Alcoholism and Inebriety: An Etiological Study," *British Journal of Inebriety* 2, no. 2 (1904), 75–76; Jones, "Relation of Inebriety to Mental Disease," 5–6; British Medical Association, "The Effects of Alcohol," *Lancet,* August 8, 1891, 310.

For examples from France, see "De l'alcoolisme: travaux recents sur ce sujet," *Annales d'Hygiène Publique*, 2nd ser., 15 (1861), 212–218; "Considérations générales sur l'alcoolisme, et plus particulièrement des effets toxiques produit sur l'homme par la liqueur d'absinthe," *Annales Médico-psychologiques*, 3rd ser., 6 (1860), 634–637; Auguste Voisin, "De l'état mental dans l'alcoolisme aigue et chronique," *Annales Médico-psychologiques*, 4th ser., 3 (1864), 5, and 4th ser., 4 (1864) 1, 16, 55; L. Lunier, "Du rôle que jouent les boissons alcooliques dans l'augmentation du nombre des cas de folie et de suicide," *Annales Médico-psychologiques*, 5th ser., 7 (1872), 321–358; A. Foville, review of V. Magnan, "De l'alcoolisme; des diverses formes du delire alcoolique et de leur traitement," *Annales Médico-psychologiques*, 5th ser., 11 (1874), 488–493; Morel, *Traité des dégénérescences*, 80.

43. Rush's thoughts on alcohol's habit-forming potential are quoted in Sournia, *History of Alcoholism*, 29. Spode, "Transubstantiations of the Mystery," 125; Kerr, *Inebriety*, 3; Frederic C. Coley, "Some Points in the Etiology of Inebriety," *British Journal of Inebriety* 2, no. 1 (1904), 24–25; "The Pathology of Inebriety," *Lancet*, April 16, 1887, 784; T. Claye Shaw, "The Psychology of the Inebriate," *British Journal of Inebriety* 3, no. 2 (1905), 62.

44. Kerr, "Alcoholism and Drug Habits," 5; Voisin, "De l'état mental dans l'alcoolisme," 2–3, 7, 65–66; George Archdall Reid, "Human Evolution and Alcohol," *British Journal of Inebriety* 1, no. 3 (1904), 193; Hehir, *Opium*, 223, 232–235, 292–293; Coley, "Some Points in the Etiology of Inebriety," 26.

45. Crothers, "Alcoholism and Inebriety," 70.

46. Richard Davenport-Hines, *The Pursuit of Oblivion: A Global History of Narcotics, 1500–2000* (London: Weidenfeld & Nicolson, 2001), 11–12, 14, 20, Sydenham quoted on p. 14; Baron de Tott, *Mémoires du Baron de Tott*, 87; *Lancet*, "Opium Eating," 685; Yvorel, *Les poisons de l'esprit*, 50–51, 67–75; Crumpe, *Inquiry into Nature and Properties*, 45; Berridge, *Opium and the People*, 77–78; Hayter, *Opium and the Romantic Imagination*, 34.

47. On Crabbe and Coleridge's opium habits, see Milner, *L'imaginaire des drogues*, 25, 27; and Jay, *Emperors of Dreams*, 63. De Quincey, "Confessions of an English Opium Eater," 58–61, 69–71, 86–87; Paul-Émile Botta, *De l'usage de fumer l'opium* (Paris: Imprimerie de Didot le Jeune, 1829), 15; Sigmond, "Lectures on Materia Medica, VI," 355, 357; *Lancet*, "Opium Habit," 692; "Nerve Exhaustion and Opium," *Lancet*, October 12, 1889, 754; "The Prescription of Opium Smoking," *Lancet*, May 24, 1890, 1138; A. S. Thelwall, *The Iniquities of the Opium Trade with China: Being a Development of the Main Causes Which Exclude the Merchants of Great Britain from the Advantages of an Unrestricted Commercial Intercourse with That Vast Empire* (London: Wm. H. Allen and Co., 1839), 22.

48. Little, "On the Habitual Use of Opium," 528–529. Little's work provided most of the evidence on opium that was cited in Morel's *Traité des dégénérescences*, one of the most influential sociomedical treatises in France in the nineteenth century. Morel, *Traité des dégénérescences*, 167.

49. Little, "On the Habitual Use of Opium," 528–529; Morel, *Traité des dégénérescences*, 167.

50. Morel, *Traité des dégénérescences*, 142, 164; M. le Comte d'Escayrac de Lauture, *Le désert et le Soudan* (Paris: Librairie de J. Dumaine, 1853), 221; Fonssagrives, "Opium," 247; G. Morache, "Pékin et ses habitants," *Annales d'Hygiène Publique*, 2nd ser., 32 (1869), 316–317; W. J. Moore, *The Other Side of the Opium Question* (London: J. & A. Churchill, 1882), 29, 36; Kerr, "Alcohol and Drug Habits," 73, 82; G. H. Smith, "On Opium-Smoking among the Chinese," *Lancet*, Feb-

ruary 19, 1842, 709; "Review of 'Confessions of an Opium-Eater: New Edition,'" *Lancet*, July 16, 1853, 58; Hehir, *Opium*, 121; "First Report of the Royal Commission on Opium, with Minutes of Evidence and Appendices," *Parliamentary Papers: Accounts and Papers* 60 (1894), 583–766, testimony of Rev. W. H. Collins on p. 597.

51. Moore, *Other Side of the Opium Question*, 36, 39, 76; Hehir, *Opium*, 112; Kerr, "Alcohol and Drug Habits," 72–76; Kerr, *Inebriety*, 102; Morel, *Traité des dégénérescences*, 164; "Proceedings of Westminster Medical Society, December 21, 1839," *Lancet*, January 4, 1840, 544; Smith, "On Opium-Smoking among the Chinese," 709; D. H. Cullmore, "Opium Traffic," *British Medical Journal*, June 25, 1881, 1025; "Morphiinomania [*sic*]," *Lancet*, September 10, 1887, 532; I. Pidduck, "Opium-Taking," *Lancet*, July 5, 1851, 21.

52. Yvorel, *Les poisons de l'esprit*, 115; "Hospital Facts and Observations, Illustrative of the Efficacy of the New Remedies," *Lancet*, June 5, 1830, 393; George G. Sigmond, "Lectures on Materia Medica and Therapeutics, Lecture IX," *Lancet*, January 21, 1837, 636; R. H. Bakewell, "Hypodermic Injection of Morphia," *Lancet*, July 14, 1860, 47; Jean-Baptiste Fonssagrives, "Morphine et amorphine," in Dechambre, *Dictionnaire encyclopédique des sciences médicales*, 509; Benjamin Ball, *La morphinomanie* (Paris: Asselin et Hoozear, 1885), 5; Liedekerke, *La belle époque de l'opium*, 35; Edward Levinstein (trans. Charles Harrer), *Morbid Craving for Morphia (Die Morphiumsucht)* (1877) (London: Smith Elder & Co., 1878), 1; Clifford Allbutt, "On the Abuse of Hypodermic Injections of Morphia," *The Practitioner* 5 (Dec. 1870), 327.

53. Allbutt, "On the Abuse of Hypodermic Injections," 327; Daniel Jouet, *Étude sur le morphinisme chronique* (Paris: Alphonse Derenne, 1883), 17–18, 25; Jules Claretié, "La vie à Paris," *Le Temps*, October 4, 1881; Milner, *L'imaginaire des drogues*, 208; Levinstein, *Morbid Craving for Morphia*, 2; B. Ball and O. Jennings, "Considérations sur le traitement de la morphinomanie," *Bulletin de l'Académie de Médecine* 17 (1887), 373; *Lancet*, "Morphiinomania," 532; Liedekerke, *La belle époque de l'opium*, 106, 109–114; Fonssagrives, "Opium," 253; Boon, *Road of Excess*, 48; "The Opium Habit in Europe," *Lancet*, September 22, 1894, 698; "Morphia Disease," *Lancet*, April 15, 1876, 578; "The Paris Universal Exhibition," *Lancet*, August 31, 1889, 451; "The Supply of Morphine to Morphinomaniacs," *Lancet*, April 4, 1891, 803.

54. Claudius Gaudry, *Du morphinisme chronique et de la responsabilité pénale chez les morphinomanes* (Coulommiers, France: P. Brodard et Gallois, 1886), 11, 24–25; Jouet, *Étude sur le morphinisme chronique*, 17, 25. Some writers believed that morphine intoxication was similar to alcoholic drunkenness. See, for example, Levinstein, *Morbid Craving for Morphia*, 4. Most others, however, maintained that the morphine experience was like that of taking opium. For examples of British medical literature emphasizing the commonalities between morphine and opium intoxication, see *Lancet*, "Morphia Disease," 578; John Harley, *The Old Vegetable Neurotics: Hemlock, Opium, Belladonna, and Henbane* (London: Macmillan and Co., 1869); John Simon, "A Course of Lectures in General Pathology: Delivered at St. Thomas's Hospital, Lecture X," *Lancet*, August 10, 1850, 170; S. A. K. Strahan, "Treatment of Morphia Habitués by Suddenly Discontinuing the Drug," *Lancet*, March 29, 1884, 561–562; Benjamin Ward Richardson, "Morphia Habitués," *Lancet*, December 15, 1883, 1046. For examples from France, see Ball, *La morphinomanie*, 18–20; Georges Pichon, *Le morphinisme: impulsions délictueuse, troubles physiques et mentaux des morphinomanes—leur capacité et leur situation juridique—cause, déontologie et prophylaxie du vice morphinique* (Paris: Octave Doin, 1889), 35–37; Paul Garnier, "De l'état mental et de la responsabilité pénale dans le morphionisme [*sic*]," *Annales Médico-psychologiques*,

7th ser., 3 (1886), 359; Henri Guimbail, *Les morphinomanes* (Paris: J. B. Baillière, 1891), 18; Dr. Motet, "La morphinomanie," *Annales d'Hygiène Publique*, 3rd ser., 10 (1883), 30–31; Kerr, "Alcoholism and Drug Habits," 71.

55. Allbutt, "On the Abuse of Hypodermic Injections," 327–330.

56. Levinstein, *Morbid Craving for Morphia*, 3–4; E. Levinstein, *La morphiomanie, monographie basée, sur des observations personnelles* (Paris: Masson, 1878).

57. Levinstein, *Morbid Craving for Morphia*, 4–5, 15–16, 40–41, 51–52, 84; Pichon, *Le morphinisme*, 4, 42, 168, 189–190, 412; Guimbail, *Les morphinomanes*, 6; E. Marandon de Montyel, "Contribution à l'étude de la morphinomanie," *Annales Médico-psychologiques*, 7th ser., 1 (1885), 46; Strahan, "Treatment of Morphia Habitués," 562; Paul Garnier, "Morphinisme: avec attaques hysterio-epileptiques causes par l'abstinence de la dose habituelle du poison: vol à l'étalage," *Annales d'Hygiène Publique*, 3rd ser., 15 (1886), 314; Richardson, "Morphia Habitués," 1045; Henri Guimbail, "Crimes et délits commis par les morphinomanes," *Annales d'Hygiène Publique*, 3rd ser., 26 (1891), 484, 496; Ball, *La morphinomanie*, 54; Kerr, "Alcohol and Drug Habits," 83; Motet, "La morphinomanie," 28–31; Kerr, *Inebriety*, 104–105; Jouet, *Étude sur le morphinisme chronique*, 64; Gaudry, *Du morphinisme chronique*, 8, 9, 18, 25, 72–73, 496.

58. Jouet, *Étude sur le morphinisme chronique*, 23, 26; Levinstein, *Morbid Craving for Morphia*, 5, 13; Ball, *La morphinomanie*, 22–23, 42; Richardson, "Morphia Habitués," 1045–1046.

59. Liedekerke, *La belle époque de l'opium*, 113; Strahan, "Treatment of Morphia Habitués," 562; Guimbail, *Les morphinomanes*, 43, 86–87; Jouet, *Étude sur le morphinisme chronique*, 18; F. Senlecq, "Un cas de morphinomanie," *Annales Médico-psychologiques*, 8th ser., 1 (1895), 34; Ball, *La morphinomanie*, 9.

60. Levinstein, *Morbid Craving for Morphia*, 5–6, 8. Not all medical researchers believed that morphine had no negative effect on cognitive or intellectual functioning. Motet, for example, wrote that "even if physical health has not been altered, this is not necessarily the case for mental health. The disorder on this front is profound, and such that there is reason to fear that the cure will be incomplete, even after special treatment to break . . . the habit." Motet, "La morphinomanie," 32. However, many doctors thought that mental capacities remained fully intact, even though the "moral sense" did not, during the drug-using career of a morphine addict. See, for example, Ball, *La morphinomanie*, 13; Pichon, *Le morphinisme*, 26; Kerr, "Alcohol and Drug Habits," 73; Guimbail, "Crimes et délits," 487–488, 492; *Lancet*, "Morphia Disease," 579; Richardson, "Morphia Habitués," 1046; Gaudry, *Du morphinisme chronique*, 73. For a better understanding of how, according to doctors, the "moral sense" could become compromised without other areas of mental functioning being debilitated, see Philippe Joseph Boileau de Castelnau, "Des maladies du sens moral," *Annales Médico-psychologiques*, 3rd ser., 6 (1860), 350–376.

61. Berridge, *Opium and the People*, 144–145; Ball, *La morphinomanie*, 4; Jouet, *Étude sur le morphinisme chronique*, 19, 65; Parssinen, *Secret Passions, Secret Remedies*, 83–84, 94, 104; Ernest Chambard, *Les morphinomanes: étude clinique, médico-légale et thérapeutique* (Paris: Rueff, 1893), 8; Pichon, *Le morphinisme*, 2; "Chronique," *Annales Médico-psychologiques*, 6th ser., 10 (1883), 356; Levinstein, *Morbid Craving for Morphia*, 3; Guimbail, *Les morphinomanes*, 30.

62. Yvorel, *Les poisons de l'esprit*, 117–119; Ball, *La morphinomanie*, 15, 36; Gaudry, *Du morphinisme chronique*, 24; Société médico-psychologique, "Séance du 26 novembre, 1888," *Annales Médico-psychologiques*, 7th ser., 8 (1888), 145; Motet, "La morphinomanie," 26; Marandon de Montyel, "Contribution," 50; Richardson, "Morphia Habitués," 1046; Strahan, "Treatment of

Morphia Habitués," 561; Brouardel, *Opium, morphine, et cocaïne*, 141–142; Pichon, *Le morphinisme*, v, 7, 33, 37; *Lancet*, "Morphiinomania," 532; Guimbail, *Les morphinomanes*, 33.

63. Guimbail, "Crimes et délits," 495–496.

64. Ibid.; Berridge, *Opium and the People*, 155–157; Fonssagrives, "Opium," 254; Motet, "Sur le concours pour le Prix Falret en 1896," *Bulletin de l'académie de Medicine*, 3rd ser., 36 (1896), 531; Marandon de Montyel, "Contribution," 62; Hayter, *Opium and the Romantic Imagination*, 39; Jouet, *Étude sur le morphinisme*, 19; *Lancet*, "Opium Eating," 686; Kerr, "Alcohol and Drug Habits," 79; Guimbail, *Les morphinomanes*, 10, 73–75; Pichon, *Le morphinisme*, 34; Ball, *La morphinomanie*, 22.

65. Susan Sontag, *Illness as Metaphor* (New York: Farrar, Straus and Giroux, 1977), 21, 85; Michel Foucault, *Madness and Civilization: A History of Insanity in the Age of Reason* (London: Routledge, 2001); Peter Conrad and Joseph W. Schneider, *Deviance and Medicalization: From Badness to Sickness* (Philadelphia: Temple University Press, 1992), 30–35; Peter Conrad, "Medicalization and Social Control," *Annual Review of Sociology* 18 (1992), 211–213.

66. J. F. B. Tingling, *The Poppy Plague and England's Crime* (London: Elliot Stock, 1876), 17; Thelwall, *Iniquities of the Opium Trade*, 8; Farrère, *Black Opium*, 145.

67. Baudelaire, *Artificial Paradises*, 60; Benedict Anderson, *Imagined Communities: Reflections on the Origin and Spread of Nationalism* (London: Verso, 1991), 6–7; Derek Heater, *What Is Citizenship?* (Cambridge: Polity Press, 1999), 56, 64.

68. Dipak K. Gupta, *Path to Collective Madness: A Study in Social Order and Political Pathology* (Westport, CT: Praeger, 2001), 72–73; Ehrenberg, *L'individu incertain*, 21, 27.

Chapter Two • Anti-narcotic Nationalism

Epigraph. A Competition Wallah, "Letters on India: Letter III about Opium and Other Things," *Macmillan's Magazine* 8 (May/Oct. 1863), 277.

1. John Stuart Mill, *On Liberty, with Subjection of Women and Chapters on Socialism* (1869) (Cambridge: Cambridge University Press, 1989), 13; James B. Bakalar and Lester Grinspoon, *Drug Control in a Free Society* (Cambridge: Cambridge University Press, 1984), 2–34; Robert J. MacCoun and Peter Reuter, *Drug War Heresies: Learning from Other Vices, Times, and Places* (Cambridge: Cambridge University Press, 2001), 55–72; Alain Ehrenberg, *L'individu incertain* (Paris: Calmann-Levy, 1995), 27, 37–38, 42, 67, 70.

2. Virginia Berridge, *Opium and the People: Opiate Use and Drug Control Policy in Nineteenth and Early Twentieth Century England* (London: Free Association Books, 1999), 34–35, 148, 225–226; "The Pharmacy Act, 1868," sects. 15 and 17, *Halsbury's Statutes of England* (London: Butterworth & Co., 1930), 11: 689–690; Norman Kerr, *Inebriety: Its Etiology, Pathology, Treatment, and Jurisprudence*, 2nd ed. (London: H. K. Lewis, 1889), 109.

3. Berridge, *Opium and the People*, 226, 296; Robert Armstrong-Jones, "Drug Addiction in Relation to Mental Disorder," *British Journal of Inebriety* 12, no. 3 (1915), 128; Terry M. Parssinen, *Secret Passions, Secret Remedies: Narcotic Drugs in British Society, 1820–1930* (Philadelphia: Institute for the Study of Human Issues, 1983), 104; S. A. K. Strahan, "Treatment of Morphia Habitués by Suddenly Discontinuing the Drug," *Lancet*, March 29, 1884, 561; Kerr, *Inebriety*, 116; "Morphiinomania [*sic*]," *Lancet*, September 10, 1887, 532.

4. Parssinen, *Secret Passions, Secret Remedies*, 83–85; Seymour J. Sharkey, "The Treatment of Morphia Habitués by Suddenly Discontinuing the Drug," *Lancet*, December 29, 1883, 1120.

5. Parssinen, *Secret Passions, Secret Remedies*, 84; Berridge, *Opium and the People*, 149, 228.

6. For an overview of how the British political order was imagined during this period, see Frank Dobbin, *Forging Industrial Policy: The United States, Britain, and France in the Railway Age* (Cambridge: Cambridge University Press, 1994), 2, 20–21, 161–164; Brian S. Turner, "Outline of a Theory of Citizenship," *Sociology* 24, no. 2 (May 1990), 207–208; Philip Harling, "The Powers of the Victorian State," in Peter Mandler (ed.), *Liberty and Authority in Victorian Britain* (Oxford: Oxford University Press, 2006), 26–27, 30–31, 45; Martin Pugh, *State and Society: A Social and Political History of Britain since 1870* (London: Hodder Education, 2008), 52–54, 64; Linda Colley, *Britons: Forging the Nation, 1707–1837* (New Haven, CT: Yale University Press, 1992), 36–37; Brian Harrison, *The Transformation of British Politics, 1860–1995* (Oxford: Oxford University Press, 1996), 16–17, 25, 115; Boyd Hilton, *A Mad, Bad, and Dangerous People? England 1783–1846* (Oxford: Clarendon Press, 2006), 636–637.

On the relationship between British thinking about the nation and economic policy, see Pugh, *State and Society*, 10–11, 49–51, 57–58; Harling, "Powers of the Victorian State," 26; Dobbin, *Forging Industrial Policy*, 25, 162; J. P. Parry, "Liberalism and Liberty," in Mandler, *Liberty and Authority*, 71; Paul Johnson, "Market Disciplines," in Mandler, *Liberty and Authority*, 211; Daniel Verdier, *Democracy and International Trade: Britain, France, and the United States, 1860–1990* (Princeton, NJ: Princeton University Press, 1994), 138–145.

For examples of how interventionist policies were implemented to preserve or enhance individualism, see Peter Mandler, "Introduction: State and Society in Victorian Britain," in Mandler, *Liberty and Authority*, 1–2; G. R. Searle, *Morality and the Market in Victorian Britain* (Oxford: Clarendon Press, 1998), 265, 274; Colley, *Britons*, 56, 66–67; Perry Anderson, *Lineages of the Absolutist State* (London: Verso, 1979), 134–135; Peter Baldwin, "The Victorian State in Comparative Perspective," in Mandler, *Liberty and Authority*, 65; Harling, "Powers of the Victorian State," 26; Parry, "Liberalism and Liberty," 85; Eugenio F. Biagini, "Introduction: Citizenship, Liberty, and Community," in Eugenio F. Biagini (ed.), *Citizenship and Community: Liberals, Radicals, and Collective Identities in the British Isles, 1865–1931* (Cambridge: Cambridge University Press, 1996), 1–3, 9; Joanna Innes, "Forms of 'government growth,' 1780–1830," in David Feldman and Jon Lawrence (eds.), *Structures and Transformations in Modern British History* (Cambridge: Cambridge University Press, 2011), 98–99.

7. Parry, "Liberalism and Liberty," 71, 73, 78, 99; Searle, *Morality and the Market*, 3, 198–201, 265; Anthony Howe, "Towards the 'Hungry Forties': Free Trade in Britain, c. 1880–1906," in Biagini, *Citizenship and Community*, 194; Frank Trentmann, "The Strange Death of Free Trade: The Erosion of 'Liberal Consensus' in Great Britain, c. 1903–1932," in Biagini, *Citizenship and Community*, 221–222, 249; Liah Greenfeld, *Nationalism: Five Roads to Modernity* (Cambridge, MA: Harvard University Press), 86; Susan Pedersen, *Family, Dependence, and the Origins of the Welfare State* (Cambridge: Cambridge University Press, 1993), 35; Bill Schwarz, "Night Battles: Hooligan and Citizen," in Mica Nava and Alan O'Shea (eds.), *Modern Times: Reflections on a Century of English Modernity* (London: Routledge, 1996), 118; Joanna Innes and Arthur Burns, "Introduction," in Arthur Burns and Joanna Innes (eds.), *Rethinking the Age of Reform, Britain 1780–1850* (Cambridge: Cambridge University Press, 2003), 43; Colley, *Britons*, 5–6, 11, 17, 52.

8. Ian Baucom, *Out of Place: Englishness, Empire, and the Locations of Identity* (Princeton,

NJ: Princeton University Press, 1999), 12; Cannon Schmitt, *Alien Nation: Nineteenth-Century Gothic Fictions and English Nationality* (Philadelphia: University of Pennsylvania Press, 1997), 2; Robert Mudie, *China and Its Resources, and Peculiarities, Physical, Political, Social, and Commercial: With a View of the Opium Question, and a Notice of Assam* (London: Grattan and Gilbert, 1840), 178; Hunt Jenin, *The India-China Opium Trade in the Nineteenth Century* (Jefferson, NC: McFarland & Co., 1999), 102–103; Mike Jay, *Emperors of Dreams: Drugs in the Nineteenth Century* (Sawtry, UK: Dedalus, 2000), 71; Barry Milligan, *Pleasures and Pains: Opium and the Orient in Nineteenth-Century British Culture* (Charlottesville: University of Virginia Press, 1995), 19, 29–30, 84–85.

9. "Opium Smoking at the East End of London," *Daily News*, 1864, reprinted in Hartman Henry Sulzberger (ed.), *All about Opium* (London: Wertheimer, Lea and Co., 1884), 175–177.

10. Charles Dickens, "Lazarus, Lotus Eating," *All the Year Round*, May 12, 1866, 421–425.

11. Parssinen, *Secret Passions, Secret Remedies*, 52–53; Mary Roth, "Victorian Highs: Detection, Drugs, and Empire," in Janet Farrell Brodie and Marc Redfield (eds.), *High Anxieties: Cultural Studies in Addiction* (Berkeley: University of California Press, 2002), 87, 91; Milligan, *Pleasures and Pains*, 13, 85–86, 101. The fear of Asians' use of narcotics to destroy British virtues continued past the texts discussed here and survived into the twentieth century. See Parssinen, *Secret Passions, Secret Remedies*, 117–121.

12. Oscar Wilde, "The Picture of Dorian Gray" (1890), in Oscar Wilde, *The Complete Plays, Poems, Novels, and Stories of Oscar Wilde* (London: Magpie, 1993), 142; Arthur Conan Doyle, "Man with the Twisted Lip" (1891), in Arthur Conan Doyle (ed. Leslie S. Klinger), *The New Annotated Sherlock Holmes* (New York: W. W. Norton & Co., 2005), 1: 162, 167; "A Night in an Opium Den," *Strand Magazine* 1 (1891), 625. Similar descriptions infiltrated the medical press as well. See, for example, "Opium-Smoking," *Lancet*, October 26, 1872, 613–614.

13. Wilde, "Picture of Dorian Gray," 142–143; Conan Doyle, "Man with the Twisted Lip," 163; *Strand*, "Night in an Opium Den," 626–627.

14. Conan Doyle, "Man with the Twisted Lip," 159; *Strand*, "Night in an Opium Den," 625.

15. Conan Doyle, "Man with the Twisted Lip," 191–192.

16. A. S. Thelwall, *The Iniquities of the Opium Trade with China: Being a Development of the Main Causes Which Exclude the Merchants of Great Britain from the Advantages of an Unrestricted Commercial Intercourse with that Vast Empire* (London: Wm. H. Allen and Co., 1839); John F. Richards, "Opium and the British Indian Empire: The Royal Commission of 1895," *Modern Asian Studies* 36, no. 2 (2002), 382–383; Berridge, *Opium and the People*, 174–176, 189.

17. "Letter from Storrs Turner, of the Society for the Suppression of the Opium Trade to H. Henry Sultzberger, June 10 1882," in Sulzberger, *All about Opium*, 96; Tong Kingsing, "Tong Kingsing on the Sincerity of the Chinese Government," in Sultzberger, *All about Opium*, 123; *Hansard Parliamentary Debates*, House of Commons, 3rd ser., vol. 352, cols. 286–287 (Apr. 10, 1891); "The Anti-Opium Meeting at the Mansion House," in Sultzberger, *All about Opium*, 2–15; Richards, "Opium and the British," 383.

18. "Anti-Opium Meeting at the Mansion House," 3, 16.

19. Richards, "Opium and the British," 382; Rutherford Alcock, "The Opium Trade," in Sultzberger, *All about Opium*, 28; "Sir George Birdwood's First Letter to 'The Times,'" in Sultzberger, *All about Opium*, 22–23; W. J. Moore, *The Other Side of the Opium Question* (London: J. & A. Churchill, 1882), 47; "Sir George Birdwood's Second Letter to 'The Times,'" in Sultzberger,

All about Opium, 81, 83; "A Defence of Opium Smoking," in Sultzberger, *All about Opium*, 164; "Two Rival Truths about Opium," in Sultzberger, *All about Opium*, 183; Patrick Hehir, *Opium: Its Physical, Moral, and Social Effects* (London: Baillière, Tindall and Co., 1895), ix, 12, 205.

20. Hehir, *Opium*, vii–x, 5–6, 205, 212, 235; Alcock, "Opium Trade," 53–54, 62–64; "Sir George Birdwood's First Letter," 22, 24; "Defence of Opium Smoking," 164; Dr. Ayres, "On Opium Smoking," in Sultzberger, *All about Opium*, 165; Moore, *Other Side*, 80; "Two Rival Truths about Opium," 185, emphasis in original.

21. Richards, "Opium and the British," 382, 386; Royal Commission on Opium, *Final Report* (London: Her Majesty's Stationery Office, 1895), 1; Berridge, *Opium and the People*, 187.

22. Royal Commission on Opium, *Final Report*, 4, 14–15; Richards, "Opium and the British," 389–392, 395.

23. "First Report of the Royal Commission on Opium, with Minutes of Evidence and Appendices," *Parliamentary Papers: Accounts and Papers* 60 (1894), 596–607, 622, 626–630.

24. "First Report of the Royal Commission," 652–653, 661–670, 691, 701, 706; "Royal Commission on Opium: Minutes of Evidence, November 18–December 29 1893, with Appendices, Volume II," *Parliamentary Papers: Accounts and Papers* 61 (1894), 64, 75, 85, 106, 201, quoted testimony of Surgeon-Lieutenant-Colonel O'Brien on p. 82.

25. "First Report of the Royal Commission," 652, 661–670, 692; "Royal Commission on Opium," 31, 85, 135, quoted testimony of John Lambert on p. 131.

26. "First Report of the Royal Commission," 674, 699, 721.

27. Royal Commission on Opium, *Final Report*, 15–22, 59, 67, 93–95.

28. Many commentators and historians have dismissed the 1895 Royal Commission as a "whitewash," its work undermined by the machinations of government administrators in India. See, for example, Berridge, *Opium and the People*, 187; Jasper Woodcock, "Commissions (Royal and Other) on Drug Misuse: Who Needs Them?" *Addiction* 90, no. 10 (1995), 1299. Yet, though there may have been a vested government interest in the eventual outcomes, a close reading of the transcripts reveals that the commission members, as historian John F. Richards concludes, "faithfully followed their Parliamentary instructions, reported accurately and drew reasonable conclusions from their witnesses and evidence." Richards, "Opium and the British," 380.

29. Berridge, *Opium and the People*, 202; Parssinen, *Secret Passions, Secret Remedies*, 116–121; Louise Foxcroft, *The Making of Addiction: The "Use and Abuse" of Opium in Nineteenth-Century Britain* (Aldershot, UK: Ashgate, 2007), 65.

30. "Rapport et Ordonnance Royale du 29 octobre 1846," in *Recueil des travaux du comité consultatif d'hygiène publique de France et des actes officiels de l'Administration sanitaire* (Paris: Librairie J. B. Baillière & fils, 1873), 2: 339, 342–344; "Décret du 8 juillet 1850 concernant la vente des substances vénéneuses," in *Recueil des travaux*, 348–349; Jean-Jacques Yvorel, *Les poisons de l'esprit: drogues et drogués au XIXᵉ siècle* (Paris: Quai Voltaire, 1992), 100–106.

31. In 1907, Britain imported 453,015 more pounds of raw opium than it reexported, whereas France imported only 11,656 more pounds than it reexported. In 1911, the population of Great Britain was approximately 36 million, and that of France was about 40 million. Based on these numbers, I calculate that per capita opium consumption was approximately 4.3 times higher for the British than for the French (12.58 pounds per thousand population in Britain vs. 2.91 pounds per thousand in France). These numbers do not take into account the amount of opium

used to manufacture morphine and other pharmaceuticals. *Report of the International Opium Commission, Shanghai China, February 1 to February 26, 1909* (London: P. S. King & Son, 1909), 2: 122–123, 161. Population statistics are from B. R. Mitchell, *International Historical Statistics, Europe 1750–2000* (Houndmills, UK: Palgrave Macmillan 2003), 4, 8. The amount of opium in France might have been higher, however, since there was probably more undocumented smuggling of opium into France than into Britain.

On the number of addicts seen by French doctors, see Georges Pichon, *Le morphinisme: impulsions délictueuse, troubles physiques et mentaux des morphinomanes—leur capacité et leur situation juridique—cause, déontologie et prophylaxie du vice morphinique* (Paris: Octave Doin, 1889), 16; Georges Pichon, "Morphinophagie: morphinisme et diathèse, " *Annales Médico-psychologiques*, 7th ser., 17 (1893), 205; Maurice Page, "Le traitement rationnel de la demorphinisation," *Annales Médico-psychologiques*, 10th ser., 3 (1913), 532–541. On the number of addiction cases cited in British medical literature, see Parssinen, *Secret Passions, Secret Remedies*, 83–85. On the tendency of French authors to exaggerate the prevalence of addiction, see Yvorel, *Les poisons de l'esprit*, 107.

32. Pichon, "Morphinophagie," 204, emphasis in original; Pichon, *Le morphinisme*, 16, 204; Benjamin Ball, *La morphinomanie* (Paris: Asselin et Hoozear, 1885), 2; Paul Garnier, "De l'état mental et de la responsabilité pénale dans le morphionisme [*sic*]," *Annales Médico-psychologiques*, 7th ser., 3 (1886), 352; Claudius Gaudry, *Du morphinisme chronique et de la responsabilité pénale chez les morphinomanes* (Coulommiers, France: P. Broadard et Gallois, 1886), 17; P. Brouardel, *Opium, morphine, et cocaïne: intoxication aigue par l'opium, mangeurs et fumeurs d'opium, morphinomanes et cocaïnomanes* (Paris: J. B. Baillière, 1906), 140; Yvorel, *Les poisons de l'esprit*, 121, 125–137; Maurice Talmeyr, *Les possédés de la morphine* (Paris: Plon, 1892).

33. Pichon, *Le morphinisme*, 7–8; E. Marandon de Montyel, "Contribution à l'étude de la morphinomanie," *Annales Médico-psychologiques*, 7th ser., 1 (1885), 45; Jules Claretié, "La vie à Paris," *Le Temps*, October 4, 1881; "The Paris Universal Exhibition," *Lancet*, August 31, 1889, 452; "The Opium Habit in Europe," *Lancet*, September 22, 1894, 698; "The Supply of Morphine to Morphinomaniacs," *Lancet*, April 4, 1891, 803; *Annales Médico-psychologiques*, "Chronique," 179; Gaudry, *Du morphinisme chronique*, 5, 17, 34, 42–54, 70–73; Garnier, "De l'état mental," 351–378; Société médico-psychologique, "Séance du 26 novembre 1888," *Annales Médico-psychologiques*, 7th ser., 8 (1888), 145; "Crimes et délits commis par les morphinomanes," *Annales Médico-psychologiques*, 7th ser., 18 (1893), 479.

34. *Report of the International Opium Commission*, 122; Chantal Descours-Gatin, *Quand l'opium finançait la colonisation en Indochine* (Paris: Éditions L'Harmattan, 1992); Dr. Lefèvre, "Les fumeurs d'opium dans la literature française contemporain," *La Revue*, October 1, 1911, 381; Docteur Ox, "Tueries d'opium," *Le Matin*, May 6, 1908; Georges Claretié, "La fumée noire," *Le Figaro*, December 7, 1907; Yvorel, *Les poisons de l'esprit*, 107–108.

35. Delphi Fabrice, *L'opium à Paris* (Paris: Felix Juven, 1907), 55; Pichon, *Le morphinisme*, 20–45; Marcel Briand and Francois Tissot, "Morphinisme familial par contagion," *Annales Médico-psychologiques*, 9th ser., 8 (1908), 124. In Britain, theories about opiate habits being contagious, though not as prevalent as in France, were sometimes mentioned. See, for example, Charles Dickens, *The Mystery of Edwin Drood* (London: Oxford University Press, 1956), 3. Ideas about the contagiousness of opiate addiction were also prevalent in the United States, particu-

larly fears about female addicts spreading addiction into the domestic sphere. See Timothy A. Hickman, *The Secret Leprosy of Modern Days: Narcotic Addiction and Cultural Crisis in the United States, 1870–1920* (Amherst: University of Massachusetts Press, 2007), 84, 88.

36. Anderson, *Lineages of the Absolutist State*, 85, 87, 101; Greenfeld, *Nationalism*, 118–122; Dobbin, *Forging Industrial Policy*, 2; Gérard Noiriel (trans. Geoffroy de Laforcade), *The French Melting Pot: Immigration, Citizenship, and National Identity* (Minneapolis: University of Minnesota Press, 1996), xvii, xx; Lloyd Kramer, "Historical Narratives and the Meaning of Nationalism," *Journal of the History of Ideas* 58, no. 3 (July 1997), 532; Gordon Wright, *France in Modern Times: From the Enlightenment to the Present* (New York: W. W. Norton & Co., 1995), 55; Turner, "Outline of a Theory of Citizenship," 209; James R. Lehning, *To Be a Citizen: The Political Culture of the Early French Third Republic* (Ithaca, NY: Cornell University Press, 2001), 8, 118; E. J. Hobsbawm, *Nations and Nationalism since 1780: Programme, Myth, Reality* (Cambridge: Cambridge University Press, 1990), 87; David A. Bell, *The Cult of the Nation in France: Inventing Nationalism, 1680–1800* (Cambridge, MA: Harvard University Press, 2001), 204–205.

37. Rogers Brubaker, *Citizenship and Nationhood in France and Germany* (Cambridge, MA: Harvard University Press, 1992), 2, 94, 101, 106; Eugen Weber, *Peasants into Frenchmen: The Modernization of Rural France, 1870–1914* (Stanford, CA: Stanford University Press, 1976), 72, 76; Hobsbawm, *Nations and Nationalism*, 87; Judith Surkis, *Sexing the Citizen: Morality and Masculinity in France, 1870–1920* (Ithaca, NY: Cornell University Press, 2006), 2.

38. Surkis, *Sexing the Citizen*, 2, 12–14; Dobbin, *Forging Industrial Policy*, 21, 23, 97–98; Jack Hayward, *Fragmented France: Two Centuries of Disputed Identity* (Oxford: Oxford University Press, 2007), 44; Wright, *France in Modern Times*, 143; Weber, *Peasants into Frenchmen*, 102–103; Peter McPhee, *A Social History of France, 1780–1880* (London: Routledge, 1992), 250–251, 259; Sanford Elwitt, *The Making of the Third Republic: Class and Politics in France, 1868–1884* (Baton Rouge: Louisiana State University Press, 1975); Brian Jenkins, *Nationalism in France: Class and Nation since 1789* (London: Routledge, 1990), 80–81, 85; Sudir Hazareesingh, *Intellectual Founders of the Republic: Five Studies in Nineteenth-Century French Republican Political Thought* (Oxford: Oxford University Press, 2001), 295, 297; Bell, *Cult of the Nation*, 207–208; Venita Datta, *Heroes and Legends of Fin-de-Siècle France: Gender, Politics, and National Identity* (Cambridge: Cambridge University Press, 2011), 3; Janet R. Horne, *A Social Laboratory for Modern France: The Musée Social and the Rise of the Welfare State* (Durham, NC: Duke University Press, 2002), 6, 25, 50, 53. For a detailed discussion of Bourgeois's theory of social solidarism, see Horne, *Social Laboratory*, 118–120.

39. On the role of education during the Third Republic in encouraging ideals of citizenship, see Weber, *Peasants into Frenchmen*, 300–338; Lehning, *To Be a Citizen*, 49; Surkis, *Sexing the Citizen*, 13; Jenkins, *Nationalism in France*, 85; Hayward, *Fragmented France*, 61. On policies concerning religious institutions, immigration, and other forms of cultural assimilation, see Weber, *Peasants into Frenchmen*, 359–364; Lehning, *To Be a Citizen*, 6, 9, 18; Brubaker, *Citizenship and Nationhood*, 106; Jeffrey Merrick and Brian T. Ragan Jr., "Introduction," in Jeffrey Merrick and Brian T. Ragan Jr. (eds.), *Homosexuality in Modern France* (New York: Oxford University Press, 1996), 3–5; Pierre Birnbaum (trans. M. B. DeBevoise), *The Idea of France* (New York: Hill and Wang, 2001), 118; Noiriel, *French Melting Pot*, xxiii, 259; Patrick Weil, *Qu'est-ce qu'un Français? Histoire de la nationalité française depuis la révolution* (Paris: Bernard Grasset, 2002), 60–61; Mary Dewhurst Lewis, *The Boundaries of the Republic: Migrant Rights and the Limits of*

Universalism in France, 1918–1940 (Stanford, CA: Stanford University Press, 2007), 13; Adrian Favell, *Philosophies of Integration: Immigration and the Idea of Citizenship in France and Britain* (New York: St. Martin's Press, 1998), 82.

On economic policy, see Dobbin, *Forging Industrial Policy*, 21; Noel Whiteside and Robert Salais, "Introduction: Political Economy and Modernisation," in Noel Whiteside and Robert Salais (eds.), *Governance, Industry, and Labour Markets in Britain and France: The Modernising State in the Mid-Twentieth Century* (London: Routledge, 1998), 4; Michel Margairaz, "Companies under Public Control in France, 1900–1950," in Whiteside and Salais, *Governance, Industry, and Labour Markets*, 27–30; Verdier, *Democracy and International Trade*, 125–135, 149.

40. Alain Ehrenberg makes a similar argument in his analysis of the French response to the drug problem in the late 1960s. See Ehrenberg, *Individu incertain*, 70–71; Alain Ehrenberg, "Comment vivre avec les drogues? Questions de recherche et enjeux politiques," *Communications* 62 (1996), 8–9.

41. Marcel-Jacques Mallat de Bassilan, *La comtesse morphine* (Paris: Bibliothèque des Deux Mondes, 1885); Jean-Louis Dubut de Laforest, *Morphine: un roman contemporain* (Librairie de la Société des Gens de Lettres, 1891), 4, 9, 12, 14, 21–22, 180–181, 191–194, 217, 219, 294.

42. Pierre Custot, *Midship* (Paris: Librairie Paul Ollendorf, 1901); Claude Farrère (trans. Samuel Putnam), *Black Opium* (1906) (New York: Nicholas L. Brown, 1929); Jules Boissière, *Fumeurs d'opium: comédiens ambulants* (Paris: Louis-Michaud, 1886).

43. Custot, *Midship*, 3–4, 11, 21–22, 25; Boissière, *Fumeurs d'opium*, 218–219.

44. Custot, *Midship*, 80, 82, 102, 108; Farrère, *Black Opium*, 143, 234, 237–238; Boissière, *Fumeurs d'opium*, 14–17, 135, 148, 151–152, 170, 221–226.

45. Custot, *Midship*, 104.

46. Boissière, *Fumeurs d'opium*, 160, 162, 170; Farrère, *Black Opium*, 145–149. "Annamites" refers to inhabitants of Annam, a region of French Indochina now part of Vietnam.

47. Boissière, *Fumeurs d'opium*, 142, 174, 180–183.

48. Ibid., 220–227.

49. Ibid., 229, 233–234, 243, 250–252, 256–257.

50. Ibid., 306.

51. Ernest Dupré, *L'affaire Ullmo: extrait des Archives d'anthropologie criminelle, de médecine légale et de psychologie normale et pathologique, No 176–177, août-septembre 1908* (Lyon: A. Rey & Cie, 1908), 7–8.

52. Ibid., 8–9; "L'affaire Ullmo: un procès de haute trahison," *Revue des Grands Procès Contemporains: Recueil d'Éloquence Judiciaire* 36 (1908), 252.

53. Dupré, *L'affaire Ullmo*, 6, 30; *Revue des Grands Procès Contemporains*, "L'affaire Ullmo," 221, 224.

54. Dupré, *L'affaire Ullmo*, 13–14; *Revue des Grands Procès Contemporains*, "L'affaire Ullmo," 214–219, 221, 224, 246–249.

55. Dupré, *L'affaire Ullmo*, 22–23, 30; *Revue des Grands Procès Contemporains*, "L'affaire Ullmo," 213, 215, 219–220, 223–224.

56. Dupré, *L'affaire Ullmo*, 29; *Revue des Grands Procès Contemporains*, "L'affaire Ullmo," 220, 223, 229, 232, 236, 245, 250; "Arrestation d'un officier de marine: chantage ou folie?" *Le Figaro*, October 25, 1907.

57. Dupré, *L'affaire Ullmo*, 10, 15, 19, 22, 26, 33–35.

58. Ibid., 19, 22, 41, emphasis added.

59. Ibid., 30–31, 36–37. Ullmo was eventually pardoned (in 1933) because many thought his original sentence too harsh. Datta, *Heroes and Legends*, 221.

60. Léon Daudet, "L'école de Judas," *Libre Parole*, October 27, 1907; "Crime contre la patrie," *Le Matin*, October 25, 1907; "L'affaire Ullmo: le traître raconte," *Le Matin*, October 30, 1907; Alfred Meynard, "L'opium," *Le Petit Marseillais*, December 23, 1907; Claretié, "La fumée noire"; "La guerre à l'opium," *Le Matin*, December 13, 1907; Docteur Ox, "Tueries d'opium"; Datta, *Heroes and Legends*, 180, 194–202, 206, 210–221, esp. 214–215.

61. For examples of anti-Semitic press coverage of the Ullmo case, see Léon Daudet, "L'opium, fait-il trahir?" *L'Action Française*, May 8, 1908; Daudet, "L'école de Judas"; Albert Monniot, "Après l'armée, la marine, après Dreyfus, Ulmo [*sic*]!" *Libre Parole*, October 26, 1907. Nevertheless, anti-Semitism played a relatively minimal role in most press reporting on the Ullmo Affair. See Datta, *Heroes and Legends*, 184.

62. Leo Gerville-Reache, "On dégrade le traître Ullmo," *Le Matin*, June 13, 1908. According to some other press accounts, Ullmo cried at the degradation, though he tried to maintain a stoic pose. See Datta, *Heroes and Legends*, 218–220.

63. In the wake of the Ullmo case, some in the press called for tighter restrictions on opium and morphine. However, in addition to opium, journalists also highlighted the role of Welsch in facilitating Ullmo's demise and argued that his appetite for drugs and his weakness for women were intertwined. See Datta, *Heroes and Legends*, 214–215.

Chapter Three • The Era of National Narcotics Control

Epigraph. Malcolm Delevingne, "Some International Aspects of the Problem of Drug Addiction," *British Journal of Inebriety* 23, no. 3 (Jan. 1935), 128.

1. *Report of the International Opium Commission, Shanghai China, February 1 to February 26 1909* (London: P. S. King & Son, 1909), 1: 26; and 2: 10–21, 78–82, 167–168, 171, 250–252; Hamilton Wright, "The International Opium Commission," *American Journal of International Law* 3, no. 4 (1909), 651; Chantal Descours-Gatin, *Quand l'opium finançait la colonisation en Indochine* (Paris: Éditions L'Harmattan, 1992), 246; William B. McAllister, *Drug Diplomacy in the Twentieth Century: An International History* (London: Routledge, 2000), 24.

2. Britain, France, Japan, Russia, Germany, Portugal, the Netherlands, Persia, China, Siam, Austria-Hungary, and Italy sent representatives to the conference. Two countries heavily involved in drug production and manufacturing that did not attend the conference were Turkey and Switzerland. McAllister, *Drug Diplomacy in the Twentieth Century*, 26–28; Richard Davenport-Hines, *The Pursuit of Oblivion: A Global History of Narcotics, 1500–2000* (London: Weidenfeld & Nicolson, 2001), 155–157; John S. Gregory, *The West and China since 1500* (Houndmills, UK: Palgrave Macmillan, 2003), 126–127; "Extracts from Report of Opium Committee of Philippine Commission," in *Patriotic Studies: Including Extracts from Bills, Acts, and Documents of United States Congress, 1888–1905* (Washington, DC: American Law, 1905), 213–231; John Palmer Gavit, *Opium* (New York: Bretano's, 1927), 14.

3. *Report of the International Opium Commission, Shanghai*, 1: 46–53, 84; McAllister, *Drug Diplomacy*, 29.

4. McAllister, *Drug Diplomacy*, 30–31; *Conference internationale de l'opium, La Haye, 1*

décembre 1911–23 janvier 1912, actes et documents (The Hague: Imprimerie Nationale, 1912), 1: 1–3, 36–38, 41, 182–184.

5. F. S. L. Lyons, *Internationalism in Europe, 1815–1914* (Leyden, Netherlands: A. W. Sythoff-Leyden, 1963), 374, 376; McAllister, *Drug Diplomacy*, 29.

6. "Opium Convention between China, France, Germany, Great Britain, Japan, the Netherlands, Persia, Portugal, Russia, Siam and the United States Signed at the Hague" (1912), in Clive Perry (ed.), *The Consolidated Treaty Series* (Dobbs Ferry, NY: Oceana Publications, 1980), 215: 300, 302; McAllister, *Drug Diplomacy*, 30–35; Lyons, *Internationalism in Europe*, 374–378; Raymond Leslie Buell, *The International Opium Conferences: With Relevant Documents* (Boston: World Peace Foundation, 1925), 115; "Opium Committee, Suggestions for Report: Second Draft," July 24, 1914, Public Record Office, British National Archives, Kew (hereafter, PRO), CUST 49/322.

7. "Opium Convention," 299–303; William J. Collins, "The Sixth Norman Kerr Lecture: The Ethics and Law of Drug and Alcohol Addiction," *British Journal of Inebriety* 13, no. 3 (Jan. 1916), 150–151.

8. McAllister, *Drug Diplomacy*, 29–37, 76, 96–97; Bertil A. Renborg, "International Control of Narcotics," *Law and Contemporary Problems* 22, no. 1 (1957), 89; Buell, *International Opium Conferences*, 73; Bertil A. Renborg, *International Drug Control: A Study of International Administration by and through the League of Nations* (Washington, DC: Carnegie Endowment for International Peace, 1943), 17–19; Catherine Carstairs, "The Stages of the International Drug Control System," *Drug and Alcohol Review* 24, no. 1 (2005), 57–59.

9. Although the 1936 Convention on Trafficking had measures calling for illicit possession to be an offense, this agreement did not take force until the fall of 1939 and had limited effects at the time, due to disruptions caused by World War II. The 1953 Opium Protocol was the first international agreement to call for restrictions on nonmedical use of opiates, and the 1961 Single Convention focused on the individual drug user, suggesting that offering and possessing drugs should also be punishable offenses. McAllister, *Drug Diplomacy*, 123; Carstairs, "Stages of the International Drug Control System," 58–61. In the 1930s, the League of Nations Health Committee established a commission to study the treatment of drug addicts, but the group's work did not lead to any international policies on the treatment of addiction. Virginia Berridge, "The Inter-war Years: A Period of Decline," *British Journal of Addiction* 85, no. 8 (1990), 1030.

10. Virginia Berridge, *Opium and the People: Opiate Use and Drug Control Policy in Nineteenth and Early Twentieth Century England* (London: Free Association Books, 1999), 120, 239; Virginia Berridge, "War Conditions and Narcotics Control: The Passing of the Defence of the Realm Act Regulation 40B," *Journal of Social Policy* 7, no. 3 (1978), 286; "The Pharmacy Act, 1868," sects. 15 and 17, in *Halsbury's Statutes of England* (London: Butterworth & Co, 1930), 11: 689–690; "The Poisons and Pharmacy Act, 1908," schedule, in *Halsbury's Statutes of England*, 11: 739; "Schedule A," in Hugh H. L. Bellot, *The Pharmacy Acts, 1851–1908* (London: Jesse Boot, 1908), 63; New Scotland Yard S.W. to Under Secretary of State, July 20, 1916, PRO, HO 45/10813/312966; Wippell Gadd, "How Far Can the Abuse of Drugs Be Prevented by Law?" *Lancet*, April 8, 1911, 933; Untitled Customs document, November 28, 1911, PRO, CUST 49/322.

11. There is evidence in medical literature hinting that some in the medical and legal communities did want stricter controls over opium during this time. For example, in 1904, a coroner suggested that "medical men ought to limit the time for which a prescription might be

used," after a woman who had been using morphine for three years died of an overdose. See "Prescriptions of Opium and Morphine," *British Medical Journal*, December 10, 1904, 1617; Virginia Berridge, "Prevention and Social Hygiene, 1900–1914," *British Journal of Addiction* 85, no. 8 (1990), 1014. However, such concerns are not prominent in the Home Office archives or in official documentation from the era. See, for example, Sir E. Grey to British Delegates to the International Opium Commission, December 21, 1908, PRO, HO 45/10500/119609; *Report of the International Opium Commission, Shanghai*, 2: 161. The London County Council's by-laws are mentioned in Terry M. Parssinen, *Secret Passions, Secret Remedies: Narcotic Drugs in British Society, 1820–1930* (Philadelphia: Institute for the Study of Human Issues, 1983), 116.

12. "Opium," no date, PRO, CUST 49/322; Untitled document from 1914, PRO, HO 45/10500/119609; "Opium Committee, Suggestions for Report, Second Draft," July 24, 1914, PRO, CUST 49/322; "Note on the Legislation Required to Give Effect to the Opium Convention in the United Kingdom," no date, PRO, CUST 49/322; Berridge, "War Conditions and Narcotics Control," 291.

13. "Opium Alleged to Be Exported—Means of Ensuring That Exportation Has Actually Taken Place," no date; "Opium," no date; both in PRO, CUST 49/322.

14. "Opium," no date; "Opium Committee, Suggestions for Report, Second Draft"; Board of Trade to Foreign Office, January 3, 1912; all in PRO, CUST 49/322. *Parliamentary Debates*, HC, 5th ser., vol. 71, col. 2169 (May 18, 1915) (hereafter, all parliamentary debates series referred to as *Hansard*; HC, House of Commons).

15. "Opium Policy Committee: Note by the Home Secretary," May 27, 1927, PRO, HO 45/20413; *Hansard*, HC, 5th ser., vol. 73, col. 2273 (July 28, 1915); Collins, "Sixth Norman Kerr Lecture," 153–154.

16. *Hansard*, HC, 5th ser., vol. 73, col. 2273 (July 28, 1915); *Hansard*, HC, 5th ser., vol. 50, cols. 214–215 (Mar. 24, 1914); Imports and Exports Office, Hong Kong, to Customs House, London, January 6, 1913, PRO, CUST 49/322. For details on Britain-based smuggling operations, see the law enforcement correspondence in PRO, HO 45/24683; Messrs. Alfred Holt & Co., "Memorial on the Subject of the Smuggling of Opium from the United Kingdom to China, the United States of America, Canada, Australia, and the Crown Colonies of the Straits Settlements and Hong Kong," February 29, 1916, PRO, PC 8/803. For an example of smuggling from Britain to the European mainland, see the case of Louis Lardenois in Archives Nationales de France, Paris (hereafter, AN), F7/14838. See also Berridge, "War Conditions and Narcotics Control," 293.

17. Holt & Co., "Memorial on the Subject of Smuggling," PRO, PC 8/803.

18. Ibid.

19. Ibid. Section 1k of the Aliens Restriction Act of 1914 empowered the government to take any action it deemed "necessary or expedient" to prosecute foreigners considered to be a threat to national interests during the war. See "The Aliens Restriction Act, 1914" (Aug. 5, 1914) and "Aliens Restriction (Amendment) Act, 1919" (Dec. 23, 1919), *Halsbury's Statutes of England* (London: Butterworth & Co, 1929), 1: 201–203. For examples of use of the Aliens Restriction Act against Chinese immigrants involved in the illicit opium trade, see Chief Constable, City Police, Glasgow, unmarked note, July 19, 1916; Inspector C.I.D. Essex Street, "Exportation of Opium—Deportation," no date; Assistant Head Constable to Undersecretary of State, April 22, 1916; "Response to the Chief Constable," no date; "Raids on Chinese Opium Houses: Chinamen to Be Deported," May 11, 1917; all in PRO, HO 45/24683.

20. Holt & Co., "Memorial on the Subject of Smuggling," PRO, PC 8/803.

21. "Restriction of Opium Traffic," June 19, 1916, PRO, PC 8/803; Colonial Office, "Conclusions Arrived at by the Interdepartmental Conference on Opium Traffic," July 15, 1916, PRO, HO 45/10500/119609.

22. On the inability of the Pharmacy Acts to control drug dealing, see "Report of Francis Carlin, Divisional Detective Inspector, Metropolitan Police, Vine Street Station, C Division Report," May 5, 1916; "Prosecution of William Charles Johnson: Improper Trafficking in Cocaine," May 15, 1916; "Judgment of Mr. Graham Campbell," May 11, 1916; Assistant Commissioner of Police, "Minutes," May 26, 1916; New Scotland Yard S.W. to Undersecretary of State, July 20, 1916; all in PRO, HO 45/10813/312966.

For a discussion of alcohol control policy during World War I, see John Greenaway, *Drink and British Politics since 1830: A Study in Policy-Making* (Houndmills, UK: Palgrave Macmillan, 2003), 91–113. On DORA regulation 40A and its shortcomings, see "The Defence of the Realm Regulations Consolidated, Reg. 40," in Charles Cook (ed.), *Manuals of Emergency Legislation: Defence of the Realm Manual,* 6th ed. (London: His Majesty's Stationery Office, 1918), 158–164; Sergeant Francis Lloyd, London District Headquarters, to Sir Edward Henry, Scotland Yard, July 20, 1916, PRO, HO 45/10813/312966; "Report of Francis Carlin," PRO, HO 45/10813/312966; "Ruthless Measures," *Daily Chronicle,* July 20, 1916, clipping in PRO, HO 45/10813/312966. For a discussion of the prosecution of cocaine dealers Horace Kingsley and Rose Edwards, see Berridge, *Opium and the People,* 249–250. For an overview of the interdepartmental committee's deliberations on Holt's memorandum, see Colonial Office, "Conclusions Arrived at by the Interdepartmental Conference," PRO, HO 45/10500/119609.

23. "The Defence of the Realm Regulations Consolidated, Reg. 51," in Cook, *Manuals of Emergency Legislation,* 184–185; "Restriction of Opium Traffic," PRO, PC 8/803.

24. "Restriction of Opium Traffic," PRO, PC 8/803; New Scotland Yard S.W. to Undersecretary of State, July 20, 1916, PRO, HO 45/10813/312966; "Conclusions Arrived at by the Interdepartmental Conference," PRO, HO 45/10500/119609.

25. "The Defence of the Realm Regulations Consolidated, Reg. 40B," in Cook, *Manuals of Emergency Legislation,* 159–162; "Order Prescribing Form of Record of Dealings in Cocaine or Opium under Regulation 40B," in Cook, *Manuals of Emergency Legislation,* 544; Malcolm Delevingne to E. C. Cunningham, August 19, 1916, PRO, HO 45/10813/312966.

26. "Defence of the Realm Regulations, 40B," 159–162.

27. Ibid.

28. "The Defence of the Realm Consolidation Act, 1914," in Cook, *Manuals of Emergency Legislation,* 5.

29. "Note to Sir E. Troup," August 18, 1916, PRO, HO 45/10813/312966.

30. Home Office Circular to Chief Constables, August 1, 1916, PRO, HO 45/10813/312966. The following documents in PRO, HO 45/144/1672/342587: Customs House, "Imports into the United Kingdom," May 30, 1918; Customs and Excise to Home Office, December 18, 1919; Customs and Excise to Home Office, February 23, 1920; "Import Restrictions: Memorandum of the Board of Trade," November 1919.

31. "Raids on Chinese Opium Houses," June 30, 1917, PRO, HO 45/24683; Home Office memorandum, April 18, 1917, PRO, HO 45/10814/312966; "Cocaine Prescriptions and Penalties," *British Medical Journal* 1, no. 1 (June 30, 1917), 896; "Cocaine," October 30, 1916, PRO,

HO 45/10813/312966; Home Office to Customs and Excise, February 12, 1920, PRO, HO 45/144/1672/342587; "Prosecution v. Anna Rose," March 8, 1921, PRO, HO 45/144/1672/342587. The findings of the 1917 Committee on the Use of Cocaine in Dentistry also indicated that the problem had never been as widespread as law enforcement officials feared. See Berridge, "War Conditions and Narcotics Control," 300–303.

32. *Hansard*, HC, ser. 5, vol. 105, col. 1829 (Apr. 19, 1918); *Hansard*, HC, ser. 5, vol. 105, col. 1138 (Apr. 5, 1918).

33. The following documents in PRO, HO 45/144/1672/342587: Metropolitan Police report, December 6, 1918; Metropolitan Police report, December 23, 1918; "Inquest on Miss Billie Carleton: Film Actor's New Evidence," *Daily Mail*, December 13, 1918, clipping; New Scotland Yard, "Statement of Mary Hicks," December 17, 1918; "Statement of Olive Richardson," December 11, 1918; Director of Criminal Investigation to Home Office, February 4, 1919; New Scotland Yard, "Statement of Lionel Herbert Belcher," December 9, 1918; "Man and Woman Arrested: Charge of Supplying Drugs to Miss Carleton," *Times* (London), December 14, 1918, clipping. See also Berridge, *Opium and the People*, 260.

34. Berridge, *Opium and the People*, 254, 260; *Hansard*, HC, ser. 5, vol. 130, col. 714 (June 10, 1920). For examples of the press reaction to the Carleton case, see the following articles in PRO, HO 45/144/1672/342587: "Victims of the Drug Habit," *Daily Express*, December 7, 1918, clipping; "The Law of Drugs," *Weekly Dispatch*, December 10, 1918; "The Cocaine Habit," *Times* (London), December 14, 1918; "Minutes: Miss Billie Carleton's Death," December 1918; "Camouflaged Drugs," *Daily Mail*, December 9, 1918, clipping; "The Drug Danger," *Daily Express*, December 9, 1918, clipping; "More Drug Victims Than Ever," *Evening News*, December 14, 1918, clipping; "Drug Peril in the West End," *Evening News*, November 30, 1918, clipping; "Price of Opium in London," *Times* (London), December 13, 1918, clipping; "Cocaine Parties," *Daily Mail*, December 14, 1918, clipping; "What Cocaine Does," *Daily Mail*, December 13, 1918, clipping; "Drug Shops," *Weekly Dispatch*, December 10, 1918, clipping; "Drug Fascination," *Daily Mail*, December 17, 1918, clipping; "Drug Taking," *Manchester Guardian*, December 17, 1918, clipping; "100 for a Tiny Bottle," *Daily Express*, December 17, 1918, clipping; "Poisons for Body and Soul," *Daily Express*, December 19, 1918, clipping; "Vice Trust in London," *Daily Express*, December 9, 1918, clipping. Some press reports argued that the extent of the opium and cocaine habits was greatly exaggerated in some newspapers. See "Dope Orgies," *Globe*, December 16, 1918, clipping in PRO, HO 45/144/1672/342587.

35. "How to Stop the Opium Traffic," *Daily Express*, December 31, 1918, clipping; *Daily Mail*, "Cocaine Parties"; *Daily Express*, "Drug Danger"; all in PRO, HO 45/144/1672/342587.

36. Metropolitan Police report, January 20, 1919; "Minutes, Home Office, Traffic in Drugs," no date; both in PRO, HO 45/144/1672/342587.

37. "Truth about the Drug Scandal," *Weekly Dispatch*, December 8, 1918, clipping; Director of Criminal Investigation to Home Office, February 4, 1919; "Minutes, Home Office, Traffic in Drugs," no date; all in PRO, HO 45/144/1672/342587.

38. "Minute Sheet: Dangerous Drugs Bill," June 3, 1920, PRO, MH 58/51.

39. On the complaints by doctors, see "Drug Inquisition: Medical Association Protest," *Times* (London), February 11, 1921, clipping in PRO, MH 58/51. On the complaints by pharmacists, see "Dangerous Drugs," February 11, 1921, PRO, MH 58/51; Colin Divall and Sean F. Johnston, *Scaling Up: The Institution of Chemical Engineers and the Rise of a New Profession*

(Dordrecht, Netherlands: Kluwer Academic Publishers, 2000), 47; Berridge, *Opium and the People*, 265. W. J. Uglow Woolcock's complaints before parliament are in *Hansard*, HC, ser. 5, vol. 130, cols. 720–721 (June 10, 1920).

40. *Hansard*, HC, ser. 5, vol. 130, cols 721–722 (June 10, 1920); Berridge, *Opium and the People*, 265–266.

41. *Hansard*, HC, ser. 5, vol. 135, col. 683 (Nov. 25, 1920); League of Nations, Advisory Committee on Traffic in Opium (hereafter, LONACTO), *Minutes of the Fourth Session Held at Geneva from January 8th to 14th 1923*, annex 3, pp. 90, 96, 106; "The Dangerous Drugs Act, 1920," *Halsbury's Statutes of England*, 11: 756–760.

42. "Dangerous Drugs Act, 1920," 761–763; LONACTO, *Minutes of the Fifth Session Held at Geneva from May 24th to June 7th 1923*, 29; Parssinen, *Secret Passions, Secret Remedies*, 139.

43. The following documents in PRO, HO 45/24842: "Dangerous Drugs: Abstract of Returns of Persons Proceeded against during the Year Ended 31st December 1921"; "Dangerous Drugs Act 1920: Abstract of Returns of Persons Proceeded against during the Year 1922 in Great Britain"; "Prosecution under the Dangerous Drugs Acts 1923: Abstract of Returns of Persons Proceeded against for Offences during the Years [*sic*] 1923"; "Dangerous Drugs Acts, 1920 and 1923: Summary of Persons and Firms Proceeded against for Offences during the Year 1924." According to H. B. Spear's study of the operation of the Dangerous Drugs Act, the numbers were slightly different: there were 445 opium cases from 1921 to 1923, and 305 investigations involving manufactured drugs during this time; in 1924, the number of drug prosecutions dropped to 98: 48 for opium and 50 for manufactured drugs. H. B. Spear, "The Growth of Heroin Addiction in the United Kingdom," *British Journal of Addiction to Alcohol and Other Drugs* 64, no. 2 (Oct. 1969), 246.

44. Spear, "Growth of Heroin Addiction," 246. All of the figures became slightly higher when cannabis was added to the list of controlled drugs in 1929. "Summary of Evidence of Dr. E. M. Niall," PRO, MH 58/278; "Minutes: Prosecutions under the Dangerous Drugs Acts during 1924," PRO, HO 45/24842; "Prosecution under the Dangerous Drugs Act," PRO, HO 45/24842; "British Trade in Dangerous Drugs," no date, PRO, HO 45/11922; Home Office to Rev. E. Dukes, November 15, 1922, PRO, HO 45/11922.

45. "Prosecution under the Dangerous Drugs Act," PRO, HO 45/24842; Stuart Anderson and Virginia Berridge, "Opium in 20th Century Britain: Pharmacists, Regulation and the People," *Addiction* 95, no. 1 (2000), 30–33; LONACTO, *Minutes of the Fourth Session*, 93.

46. Gilles Leclair, "Stupéfiants," in Maurice Aydalot, Pierre Arpaillange, and Yves Mayaud (eds.), *Dalloz répertoire de droit pénal et de procédure pénale* (Paris: Éditions Dalloz, 1997), 6: 2; "Rapport et Ordonnance Royale du 29 octobre 1846," in *Recueil des travaux du comité consultatif d'hygiène publique de France et des actes officiels de l'Administration sanitaire* (Paris: Librairie J. B. Baillière & fils, 1873), 2: 339, 342–344; "Décret du 8 juillet 1850 concernant la vente des substances vénéneuses," in *Recueil des travaux*, 348–349.

47. Jury Médicale de la Meurthe to Préfet de la Meurthe, May 4, 1853, AN, F8/240; A. J. Martin, "Importation des substances vénéneuses: question de réglementation," in *Recueil des travaux du comité consultatif d'hygiène publique de France et des actes officiels de l'Administration sanitaire* (Paris: Librairie J. B. Baillière & fils, 1887), 16: 367; P. Brouardel and J. Regnauld, "Exercice de la pharmacie: projet de révision de la loi du 21 Germinal An XI (11 avril 1803) relative à l'exercice de la pharmacie," *Recueil des travaux*, 16: 337–359; J. Regnauld,

"Exercice de la pharmacie—substances vénéneuses," *Recueil des travaux du comité consultatif d'hygiène publique de France et des actes officiels de l'Administration sanitaire* (Paris: Librairie J. B. Baillière & fils, 1892), 21: 515; Georges Pichon, *Le morphinisme* (Paris: Octave Doin, 1889), 465–466, emphasis in original.

48. Regnauld, "Exercice de la pharmacie," 514–515; Paul Brouardel, *Opium, morphine et cocaïne* (Paris: Librairie J. B. Baillière & fils, 1906), 139, 142, 144–145.

49. Procureur Général, Aix, to Ministre de la Justice, June 5, 1906, AN, BB18/2488[2].

50. Ibid. The names of the individuals involved in this case, which is documented in Ministry of Justice archives, AN, BB18/2488[2], cannot be revealed at this time. I use initials to conceal their identities.

51. The following documents in AN, BB18/2488[2] : Procureur Général, Aix, to Ministre de la Justice, June 5, 1906; Ministère de la Justice, Note, December 22, 1906; Procureur Général, Aix, to Ministre de la Justice, November 5, 1906; "Audience du 7 décembre 1906 par le Tribunal du première instances, Jugement correctionnellement dans la cause poursuivie par le Ministère public contre J.P."; "Audience du 7 décembre 1906 par le Tribunal du première instances, Jugement correctionnellement dans la case poursuivie par le Ministre public contre J.G."

52. Procureur Général, Aix, to Ministre de la Justice, July 21, 1906, AN, BB18/2488[2].

53. The following documents in AN, BB18/2488[2]: Ministère de la Justice, Note, January 9, 1907; Ministère de la Justice to Procureur Général, Rennes, January 16, 1907; Procureur Général, Rennes, to Ministre de la Justice, January 25, 1907; Procureur Général, Rennes, to Ministre de la Justice, February 25, 1907; Procureur Général, Aix, to Ministre de la Justice, December 14, 1907; Procureur Général, Aix, to Ministre de la Justice, February 23, 1908; Procureur Général, Aix, to Ministre de la Justice, April 2, 1908; Ministre de la Justice to Ministre de la Guerre, March 21, 1908; Sous-secrétaire d'État de la Guerre to Garde des Sceaux, March 4, 1908.

54. Le Président du Conseil, Ministre de l'Intérieur, to Ministre de la Justice, September 11, 1907; Ministre de la Justice to Président du Conseil, October 18, 1907; both in AN, BB18/2488[2].

55. Ministère de l'Intérieur to Ministère de la Justice, February 27, 1908, AN, BB18/2488[2].

56. Ministère de la Justice, Note, March 14, 1908; Ministère de l'Intérieur to Ministre de la Justice, April 14, 1908; both in AN, BB18/2488[2].

57. *Journal Officiel de la République Française* (hereafter, *Journal Officiel*), October 3, 1908.

58. Ibid.

59. "Circulaire—substances vénéneuses—opium," October 31, 1908, AN, BB18/2488[2].

60. Statistics on opium arrests are compiled from Ministère de la Justice, *Compte général de l'administration de la justice criminelle, 1908* (Paris: Imprimerie Nationale, 1910), 54–64; *Compte général de l'administration de la justice criminelle, 1909* (Paris: Imprimerie Nationale, 1911), 54–64; *Compte général de l'administration de la justice criminelle, 1910* (Paris: Imprimerie Nationale, 1912), 56–67; *Compte général de l'administration de la justice criminelle, 1911* (Paris: Imprimerie Nationale, 1913), 56–65; *Compte général de l'administration de la justice criminelle, 1912* (Paris: Imprimerie Nationale, 1914), 56–65; *Compte général de l'administration de la justice criminelle, 1913* (Paris: Imprimerie Nationale, 1915), 56–65.

Procureur Général, Aix, to Ministre de la Justice, May 26, 1913; Procureur Général, Paris, to Ministre de la Justice, June 17, 1913; Le Procureur de la République, Parquet du Tribunal de la Seine, to Commissaires de Police du Département de la Seine, December 31, 1912; all in AN, BB18/2488[2]. "Les fumeries d'opium sur le littoral: les perquisitions à Toulon et a Marseille," *Le*

Petit Marseillais, March 23, 1909, press clipping in AN, F7/14838; Commissaire Central, Brest, to Directeur de la Sûreté Générale, August 11, 1911, AN, F7/14838. A detective's observations on the continued prevalence of opium smoking in the military is in Notes de Ronceyaux, February 9, 14, and 25, June 26–27, 1909, AN, F7/14838.

61. Procureur Général, Paris, to Ministre de la Justice, June 17, 1913, AN, BB18/2488²; "État des poursuites exercées par le Parquet de la Seine pour infraction au décret sur l'opium," no date, AN, BB18/2488²; Unmarked police correspondence, July 15, 1909, AN, F7/14835; "Tribunal correctionnel de la Seine, Jugement du 29 novembre 1910," in *Recueil des actes officiels et documents intéressant l'hygiène publique: travaux du Conseil Supérieur d'hygiène publique de France* (Paris: J. B. Baillière & fils, 1910), 40: 719–720; "Un droguiste avait vendu 150 kilos d'opium," *Le Journal,* May 21, 1913, clipping in AN, F7/14843; "Les pharmaciens et la vente de la morphine," *Le Matin,* October 31, 1913; "La vente des poisons: on arête un préparateur en pharmacie," *Le Matin,* May 26, 1913; Notes de Ronceyaux, May 7, 1909, AN, F7/14835.

62. Procureur Général, Paris, to Ministre de la Justice, June 17, 1913, AN, BB18/2488².

63. Igor Charras, "L'état et les 'stupéfiants': archéologie d'une politique publique répressive," *Les Cahiers de la Sécurité Intérieure* 32, no. 2 (1998), 20–21; Emmanuelle Retaillaud-Bajac, *Les paradis perdus: drogues et usagers de drogues dans la France de l'entre-deux-guerres* (Rennes, France: Presses Universitaires de Rennes, 2009), 232–233; Ministère de l'Intérieur, Direction de la Sûreté Générale, Note, August 30, 1912, AN, F7/13043; Notes de Ronceyaux, February 28 and May 9, 1909, AN, F7/14835; Préfecture du Rhône to M. Sebille, June 29, 1909, AN, F7/14835. After 1914, drug cases in Paris were investigated by a vice squad that also handled cases of prostitution and pornography. See Préfet de Police to Commissaires Divisionnaires et Commissaires de Police du Ressort, "La création de la brigade des mœurs à la direction de la police judiciaires," February 11, 1914, Archives de la Préfecture, Paris, DB/411.

64. Unmarked police correspondence, July 15, 1909, AN, F7/14835; Report to Commissaire Principal, Lyon, July 16, 1909, AN, F7/14835; Contrôleur Général des Services de Recherches Judiciaires to Commissaire Central de Brest, April 30, 1913, AN, F7/14843; Préfecture du Rhône to M. Sebille, June 29, 1909, AN, F7/14835; F. Rollins to Commissaire Principal de la Sûreté Générale, March 15 and 19, 1909, and July 9, 1909, AN, F7/14835; Notes de Ronceyaux, February 7–10, 14–17, 20–22, 24–25, 27, March 2, 11, 13, 15, May 6, 9, 11, June 21, 23, 26–27, 29, and July 1, 2, 5–9, 1909, AN, F7/14835; Procureur Général de Rennes to Ministre de la Justice, June 14, 1913, AN, F7/14843; Ministère de l'Intérieur, Note, January 16, 1914, AN, F7/14835; Commissaire de Police Mobile to Contrôleur Général des Services de Recherches, January 16, 1914, AN, F7/14835; Commissaire Central de Brest to Contrôleur Général des Services de Recherches Judiciaires, May 2, 1913, AN, F7/14843. At times, the investigators were so convincing that locals would begin to think that detectives were illicit dealers. See Inspecteur de Police Mobile Cosson to Contrôleur Général des Services de Recherches Judiciaires, May 15–17, 1913, AN, F7/14843. The process for gathering intelligence in Paris may have been similar, but I was unable to find any records documenting such investigations.

65. "Rapport de M. Vignolli," July 21, 1909, AN, F7/14835; Notes de Ronceyaux, March 13, April 5, and July 7, 1909, AN, F7/14835; Vignolle to Contrôleur Général des Services de Recherches Judiciaires, September 11, 1913, AN, F7/14837; Procureur Général, Rennes, to Ministère de la Justice, June 14, 1913, AN, F7/14843; Commissaire Spéciale de Police, Toulon, to Contrôleur Général des Services de Recherches Judiciaires, January 20, 1913, AN, F7/14843.

66. Commissaire Spéciale de Police, Toulon, to Contrôleur Général des Services de Recherches Judiciaires, January 20, 1913, AN, F7/14843; Commissaire de Police Mobile Vignolle to Contrôleur Général des Services de Recherches Judiciaires, August 18, 1913, AN, F7/14842; Commissaire Spécial, Cherbourg, to Contrôleur Général des Services Judiciaires, January 20, 1917, AN, F7/14838; Commissaire Central, Toulon, to Directeur de la Sûreté Générale, February 22, 1913, AN, F7/14842; Procureur Général, Rennes, to Ministre de la Justice, May 29, 1913, AN, BB18/2488²; Commissaire Spécial, Chemins de Fer, Ports, Emigration, Marseille, to Contrôleur Général des Services de Recherches Judiciaires, June 17, 1913, AN, F7/14842.

67. Le Procureur de la République, Parquet du Tribunal de la Seine, to Commissaires de Police du Département de la Seine, December 31, 1912, AN, BB18/2488²; G. Oudard to Ministre de l'Intérieur, August 1913, AN, F7/14837; Note pour M. Vignolle, August 9, 1913, AN, F7/14837; "La guerre à l'opium: un trafiquant est arrêté," *L'Eclaireur*, September 13, 1913, clipping in AN, F7/14837; "La guerre à l'opium: importante révélations de Blazy au Judge d'Instruction," *Le Petit Provençal*, September 17, 1913, clipping in AN, F7/14837; "Comment un rédacteur du 'Matin' put se rendre compte que le trafic de l'opium est courant à Paris," *Le Matin*, April 27, 1913; René Le Somptier, "La Morphine au Quartier Latin," *Action*, September 10, 1909.

68. "Le drame de la 'villa des stupéfiants,'" *Le Matin*, September 26 and 27, 1913, clippings in AN, F7/14837; Delphi Fabrice, *Opium à Paris* (Paris, 1914). Le Commissaire de la Sûreté à M. le Commissaire Centrale, Toulon, September 16, 1913; E. Rouzier-Dorcières, "Un péril national: l'opium à Toulon," *Le Matin*, September 19, 1913; "Incident: procès-verbal de la rencontre Rouzier-Dorcières-E. Rapuc," *Dernière Heure*, no date, clipping; all in AN, F7/14837.

69. Procureur Général, Aix, to Ministre de la Justice, May 26, 1913, AN, BB18/2488²; Notes de Roncevaux, February 10, 1909, AN, F7/14835; Rapport, Contrôle Général des Services de Recherches, March 15, 1910, AN, F7/14840; Rapport, Commissariat Spécial, Marseille, January 7, 1909, AN, F7/14835; Commissaire Spécial de Police de Toulon to Commissaire Principal Charge du Contrôle Général des Services de Recherches, May 11, 1912, AN, BB18/2488².

70. Ibid.

71. Notes de Roncevaux, February 10, 26, 27 and March 11, 1909, AN, F7/14835; Rapport, Commissariat Spécial, Marseille, June 12, 1913, AN, F7/14842; Commissaire Spécial de Police de Toulon to Commissaire Principal Charge du Contrôle Général des Services de Recherches, May 11, 1912, AN, BB18/2488²; Commissaire Spécial de Police de Toulon to Commissaire Principal Charge du Contrôle Général des Services de Recherches, May 11, 1912, AN, BB18/2488²; Rapport, Commissariat Spécial, Marseille, January 7, 1909, AN, F7/14835; Contrôle Général des Services de Recherches Judiciaires to M. Vignolle, August 18, 1913, AN, F7/14837. Other prewar speculations that Germans were orchestrating the drug traffic are in Préfecture du Rhône to Contrôleur Général des Services de Recherches Judiciaires, March 20, 1913, AN, F7/14843. Similar reports surfaced in the press. See, for example, "Les opiomanes Lyonnais s'approvisionnent en Allemagne," *Le Matin*, January 5. 1913, clipping in AN, F7/14843.

72. See the correspondence of dealer Yéhé-Roubin in AN, F7/14835; Notes de Roncevaux, March 30 and July 8, 1909, AN, F7/14835; "Analyse des documents découvertes et saisies au domicile de la dame Cognat Yvonne," March 12, 1917, AN, F7/14838; Procureur Général, Rennes, to Ministre de la Justice, March 21, 1913, AN, BB18/2488². For details on the case of smuggler Louis Lardenois and his gang, see the surveillance documents in AN, F7/14838.

73. "L'opium," *Le Matin*, June 2, 1913.

74. René Le Somptier, "La morphine au Quartier Latin," *Action*, September 10, 1909.

75. "La vente des poisons: on arête un préparateur en pharmacie," *Le Matin*, May 26, 1913; "Les opiomanes lyonnais s'approvisionnent en Allemagne," *Le Matin*, January 5, 1914; Paul Sollier, "La médecine et les intoxiqués," *Le Matin*, December 26, 1912.

76. *Journal Officiel*, Sénat, April 4, 1911, 350; Ministère de l'Intérieur, Direction du Contrôle et de la Comptabilité, "Propositions de loi déposées au Sénat relatives aux stupéfiants," AN, BB18/2488²; *Journal Officiel*, Chambre des Députés, May 6, 1913, 1334–1335; *Journal Officiel*, Chambre des Députés, May 6, 1913, documents parlementaires, annexes 2689 and 2692, pp. 395–396; *Journal Officiel*, Chambre des Députés, May 14, 1913, documents parlementaires, annex 2715, pp. 565–566; *Journal Officiel*, Sénat, June 26, 1913, documents parlementaires, annex 250, p. 1068; *Journal Officiel*, Sénat, July 22, 1914, documents parlementaires, annex 258, p. 140.

77. Procureur Général, Rennes, to Ministre de la Justice, May 29, 1913; Procureur Général, Aix, to Ministre de la Justice, May 26, 1913; both in AN, BB18/2488².

78. Ibid.

79. Ibid.; Ministère de la Justice, Note, no date, AN, BB18/2488².

80. Ministère de la Justice, "Note Complémentaire," November 6, 1913, AN, BB18/2488².

81. Ibid.; Ministère de l'Intérieur, Direction du Contrôle et de la Comptabilité, Cabinet du Directeur, "Propositions de loi déposés au Sénat relative aux stupéfiants," no date, AN, BB18/2488², emphasis in original.

82. Patricia E. Prestwich, *Drink and the Politics of Social Reform: Antialcoholism in France since 1870* (Palo Alto, CA: Society for the Promotion of Science and Scholarship, 1988), 128–140, 143–178.

83. *Journal Officiel*, Chambre des Députés, February 15, 1916, 192.

84. Henry Rigal, *Les stupéfiants: opium, morphine, cocaïne, éther, hachisch, etc.* (Paris: Attinger Frères, 1916), Cazeneuve quoted on p. 20, Poldès on p. 45, Dr. Bonnet on p. 15; *Journal Officiel*, Sénat, January 27, 1916, 23–24; Raphael-Georges Levy, in *Journal des Débats*, quoted in *Journal Officiel*, Sénat, June 15, 1915, documents parlementaires, annex 207, pp. 105–106.

85. Jean-Claude Farcy, *Les camps de concentration français de la première guerre mondiale, 1914–1920* (Paris: Anthropos, 1995), 1–7, 104.

86. Préfet de Police to Ministre de l'Intérieur, June 14, 1915, AN, F7/14835; "Individus évacuées du Camp Retranché de Paris pour usage ou vente des stupéfiants," November 4, 1915, AN, F7/14835; Préfet de Police to Ministre de l'Intérieur, March 18, 1919, AN, F7/14838; Leo Poldès, "L'affaire des poisons—un scandale: Nardin en liberté, un encouragement à l'assassinat," *Bonnet Rouge*, March 28, 1916; Commissaire de la Sûreté, Toulon, to Procureur de la République, Toulon, May 15, 1916, AN, F7/14842.

87. Préfet de Police to Ministre de l'Intérieur, June 14, 1915; Ministre de l'Intérieur to Monsieur le Préfet, no date; both in AN, F7/14835. Ministre de la Guerre to Ministre de l'Intérieur, July 10, 1916; Inspecteur de Police Mobile Max Cosson to Contrôleur Général des Services des Recherches Judiciaires, June 10, 1916; Unmarked document, no date; all in AN, F7/14838.

88. *Journal Officiel*, Sénat, July 22 1915, documents parlementaires, annex 258, p. 141; *Journal Officiel*, Sénat, November 16, 1915, documents parlementaires, annex 373, p. 274; *Journal Officiel*, Sénat, January 27, 1916, 23; *Journal Officiel*, Chambre des Députés, February 15, 1916, documents parlementaires, annex 1802, pp. 192–193.

89. *Journal Officiel*, Sénat, January 27, 1916, 21–22.

90. *Journal Officiel*, July 14, 1916, 6254; *Journal Officiel*, September 19, 1916, 8256–8261.

91. *Journal Officiel*, September 19, 1916, 8256–8261; *Journal Officiel*, Sénat, November 8, 1920, documents parlementaires, annex 474, p. 867; Retaillaud-Bajac, *Les paradis perdus*, 35.

92. *Journal Officiel*, September 19, 1916, 8256–8261; Charras, "L'état et les 'stupéfiants,'" 20.

93. Charras, "L'état et les 'stupéfiants,'" 15–16; "Rapport au sous-secrétaire d'état dus service de santé militaire," *Journal Officiel*, February 6, 1918, 1349–1350; "Application à l'armée de la législation nouvelle sure les substances vénéneuses," *Journal Officiel*, February 6, 1918, 1350–1352; Ministre de la Marine, "Note sur la question des stupéfiants," no date, AN, BB18/2488²; *Journal Officiel*, November 8, 1920, annex 474, p. 867; *Journal Officiel*, July 14, 1922, 7367; LONACTO, *Minutes of the Fifth Session*, 109–110; Retaillaud-Bajac, *Les paradis perdus*, 37–38.

94. Ministère de la Justice, *Compte général de l'administration de la justice criminelle, 1920* (Paris: Imprimerie Nationale, 1926), 56–66; *Compte général de l'administration de la justice criminelle, 1921* (Paris: Imprimerie Nationale, 1927), 50–60; *Compte général de l'administration de la justice criminelle, 1922* (Paris: Imprimerie Nationale, 1927); *Compte général de l'administration de la justice criminelle, 1923* (Paris: Imprimerie Nationale, 1928), 52–60; Ministre de l'Intérieur, Notice 44, January 12, 1923; Ministre de l'Intérieur, Notice 69, January 10, 1924; both in AN, F7/14836; Retaillaud-Bajac, *Les paradis perdus*, 226.

95. Retaillaud-Bajac, *Les paradis perdus*, 426–427.

96. Ibid., 249–250, 268, 271, 426–427.

97. Ibid., 237. The following documents in AN, F7/14836: Ministre de l'Intérieur to Commissaires Spéciaux des Portes du Territoire, November 30, 1921; Ministre de l'Intérieur to Commissaires Spéciaux, Frontière Franco-suisse, Franco-italienne, Commissaires Divisionnaires des 9, 10, 11 Brigades de Police Mobile, December 16, 1921; Ministre de l'Intérieur, Circulaire, June 25, 1923; Ministre de l'Intérieur, Notice 34, October 4, 1922; Ministre de l'Intérieur, Notice 40, November 29, 1922; Ministre de l'Intérieur, Notice 45, February 1, 1923. Inspecteurs de Police Mobile Michet, Malo et Pailler, to Contrôleur Général des Services de Recherches Judiciaires, November 30, 1922, AN, F7/14837; Commissaire Divisionnaire, Chef de la XIᵉ Brigade Régionale de Police Mobile to Inspecteur Général Charge des Services de Police Criminelle, April 20, 1938, AN, F7/14840.

98. Société de médecine légale de France, "L'opportunité d'une règlementation internationale de la vente de quelques substances toxiques," December 4, 1922, AN, BB18/2488². The following documents in AN, F7/14836: Ministre de l'Intérieur to Commissaires Spéciaux des Postes et Frontières d'Alsace, Lorraine, et de Suisse, September 30, 1920; Ministre de l'Intérieur to Commissaires Spéciaux des Portes et Frontières, Nord et Est, Directeur des Services Générale de Police d'Alsace et de Lorraine à Strasbourg, Commissaire Divisionnaire 17ᵉ Brigade de Police Mobile à Strasbourg, Commissaire Spécial Armée Française du Rhin, March 2, 1921; Commissaire Spéciale, Dunkerque, to Contrôleur Général des Services de Recherches Judiciaires, May 4 1923.

99. "Note pour M. le Contrôleur Général des Services de Recherches Judiciaires," November 30, 1922, AN, F7/14837; Les Inspecteurs de Police Mobile Michet, Malo et Pailler, to Contrôleur Général des Services de Recherches Judicaires, November 30, 1922, AN, F7/14837; Commissaire Divisionnaire, Chef de la XIᵉ Brigade Régionale de Police Mobile, to Inspecteur Général Chargé des Services de Police Criminelle, April 20, 1938, AN, F7/14840. Ministre de l'Intérieur, Notice 41, December 14, 1922; Ministre de l'Intérieur, Notice 39, November 22, 1922; "Notices signalant les individus se livrant habituellement au trafic des stupéfiants, 1–113"; all in AN, F7/14836.

100. Alan A. Block, "European Drug Traffic and Traffickers between the Wars: The Policy of Suppression and Its Consequences," *Journal of Social History* 23, no. 2 (1989), 319–322; LONACTO, *Minutes of the Fifth Session*, 76, 109; *Minutes of the Twelfth Session Held at Geneva from January 17th to February 2nd 1929*, 126; and *Minutes of the Thirteenth Session Held at Geneva from January 20th to February 14th 1930*, 25–26, 36, 59–62, 231. For details on the international trafficking syndicate operating through Marseille, see files on the Whifften Company and H.M.F. Humphrey in PRO, HO/45/24799, and PRO, HO/45/24800. Other discussions of trafficking through Marseille are in Confidential List No. 294 (Washington, DC), "Marseille Traffickers," December 1, 1934; Commissaire de Police Mobile Albayez to Contrôleur Général des Services de Recherches Judiciaires, August 30, 1934; both in AN, F7/14842. See also Retaillaud-Bajac, *Les paradis perdus*, 104–105.

101. LONACTO, *Minutes of the Fifth Session*, 110; Jules Gherlerter, *Les toxicomanies: étude médico-sociale* (Paris: Librairie Louis Arnette, 1929), 25, 66, 87; Jean Goudot, *Quelques considérations sur le traitement et le pronostic des toxicomanies* (Paris: Librairie E. Le François, 1936), 1; Société de médecine légale de France, "L'opportunité d'une règlementation internationale de la vente de quelques substances toxiques," AN, BB18/2488[2].

Chapter Four • Control and Its Discontents

Epigraph. Jean Cocteau (trans. Ernest Boyd), *Opium: The Diary of an Addict* (1930) (London: George Allen & Unwin, 1933), 11–12.

1. Norman Kerr, "Alcoholism and Drug Habits," in Thomas L. Stedman (ed.), *Twentieth Century Practice: An International Encyclopedia of Modern Medical Science by Leading Authorities of Europe and America* (London: Samson Low, Martson and Co., 1895), 3: 82.

2. On the use of substitutional drugs in the treatment of opiate addiction, see I. Pidduck, "Opium-Taking," *Lancet*, July 5, 1851, 21; M. Pouchet, "Présentations d'ouvrages manuscrits et imprimés," *Bulletin de L'Académie de Médecine*, 3rd ser., 65 (1911), 336–337; G. Laughton Scott, *The Morphine Habit and Its Painless Treatment* (London: H. K. Lewis & Co., 1930), 23–24; Kerr, "Alcoholism and Drug Habits," 3: 85. On the development of addiction to substitutional drugs, see Jean-Baptiste Fonssagrives, "Opium," in A. Dechambre (ed.), *Dictionnaire encyclopédique des sciences médicales* (Paris: Imprimerie A. Lahure, 1881), 16: 257; Claudius Gaudry, *Du morphinisme chronique et de la responsabilité pénale chez les morphinomanes* (Coulommiers, France: P. Brodard et Gallois, 1886), 56; E. Marandon de Montyel, "Contribution à l'étude de la morphinomanie," *Annales Médico-psychologiques*, 7th ser., 1 (1885), 63.

For more on the recognition of cocaine as an addictive substance, see Virginia Berridge, *Opium and the People: Opiate Use and Drug Control Policy in Nineteenth and Early Twentieth Century England* (London: Free Association Books, 1999), 162, 222–223; "Considérations sur la morphinomanie et son traitement," *Annales Médico-psychologiques*, 7th ser., 13 (1891), 127. For discussion of researchers recognizing heroin's addictive potential, see Jean Leynia de la Jarrige, *Héroïne-héroïnomanie* (Paris: L. Boyer, 1902); Jules Ghelerter, *Les toxicomanies: étude médico-sociale* (Paris: Librairie Louis Arnette, 1929), 46; Jean Goudot, *Quelques considérations sur le traitement et le pronostic des toxicomanies* (Paris: Librairie E. Le François, 1936), 6–7.

3. For discussions of treatment methods tried in Britain, see Berridge, *Opium and the People*, 161–162; Kerr, "Alcoholism and Drug Habits," 83–87; Patrick Hehir, *Opium: Its Physical,*

Moral, and Social Effects (London: Baillière, Tindall and Co., 1895), 637; Departmental Committee on Morphine and Heroin Addiction, *Report* (London: His Majesty's Stationery Office, 1926), 13–15; Scott, *Morphine Habit*, 26–28, 62; H. Crichton Miller, "The Treatment of Morphinomania by the 'Combined' Method," *British Medical Journal*, November 19, 1910, 1595–1597; J. W. Astley Cooper, "The Treatment of Morphinomania by the 'Combined' Method," *British Medical Journal*, December 24, 1910, 2007; John Q. Donald, "The Treatment of Morphinomania by the Combined Method," *British Medical Journal*, February 18, 1911, 401; Percy Boulton, "An Extraordinary Morphia Case," *Lancet*, March 4, 1882, 344.

For examples of treatment methods tried in France, see Fonssagrives, "Opium," 257; Gaudry, *Du morphinisme chronique*, 65–67; Daniel Jouet, *Étude sur le morphinisme chronique* (Paris: Alphonse Derenne, 1883), 16, 63; Maurice Page, "Le traitement rationnel de la démorphinisation," *Annales Médico-psychologiques*, 10th ser., 3 (1913), 537–539; Roger Dupouy, "Quelques réflexions sur le morphinomanie," *Annales Médico-psychologiques*, 12th ser., 1 (1922), 455; "Nouveau traitement de la morphinomanie," *Annales Médico-psychologiques*, 14th ser., 1 (1934), 804. Oscar Jennings's methods for treating opiate addiction are discussed in "The Experiences of a Medical Morphinist," in Scott, *Morphine Habit*, 78–92.

4. Berridge, *Opium and the People*, 163; Jouet, *Étude sur le morphinisme*, 59, 60; Aleister Crowley, *The Diary of a Drug Fiend* (1922) (York Beach, ME: Samuel Weiser, 1981), 182, 183; Donald, "Treatment of Morphinomania," 401; W. H. Willcox, "Norman Kerr Memorial Lecture on Drug Addiction Delivered before the Society for the Study of Inebriety, October 9th 1923," *British Medical Journal*, December 1, 1923, 1016; Dr. E. W. Adams, "Report upon the Results of an Enquiry into the Present Position as Regards the Treatment of Narcotic Drug Addiction," Public Record Office, British National Archives, Kew (hereafter, PRO), MH 58/275; Société médico-psychologique, "Séance du 26 novembre 1888," *Annales Médico-psychologiques*, 7th ser., 8 (1888), 152; G. Pichon, *Le morphinisme: impulsions délictueuse, troubles physiques et mentaux des morphinomanes—leur capacité et leur situation juridique—cause, déontologie et prophylaxie du vice morphinique* (Paris: Octave Doin, 1889), 187–188; Edward Levinstein (trans. Charles Harrer), *Morbid Craving for Morphia (Die Morphiumsucht)* (London: Smith Elder & Co., 1878), 113–114, 119.

5. Pichon, *Le morphinisme*, 191–192; Crowley, *Diary of a Drug Fiend*, 182–183; Paul Garnier, "De l'état mental et de la responsabilité pénale dans le morphionisme [*sic*]," *Annales Médico-psychologiques*, 7th ser., 3 (1886), 367; Levinstein, *Morbid Craving*, 114.

6. Levinstein, *Morbid Craving*, 112–115, 120; Pichon, *Le morphinisme*, 223–224; Jouet, *Étude sur le morphinisme*, 60; Société médico-psychologique, "Séance du 26 novembre 1888," 145; Departmental Committee, *Report*, 15; Kerr, "Alcoholism and Drug Habits," 83.

7. Levinstein, *Morbid Craving*, 110–111; Fonssagrives, "Opium," 257.

8. Levinstein, *Morbid Craving*, 110–111.

9. Société médico-psychologique," Séance du 26 novembre 1888," 151; Marandon de Montyel, "Contribution," 55, 63; Willcox, "Norman Kerr Memorial Lecture," 1016; Departmental Committee, *Report*, 15; Jouet, *Étude sur le morphinisme*, 6; James Braithwaite, "A Case in Which the Hypodermic Injection of Morphia Was Suddenly Discontinued after Its Use Daily in Large Doses for Seven Years," *Lancet*, December 21, 1878, 874; Seymour J. Sharkey, "The Treatment of Morphia Habitués by Suddenly Discontinuing the Drug," *Lancet*, December 29, 1883, 1121; St.

Thomas Clarke, "Treatment of the Habit of Injecting Morphia by Suddenly Discontinuing the Drug," *Lancet*, September 20, 1884, 491; Crowley, *Diary of a Drug Fiend*, 183.

10. Crowley, *Diary of a Drug Fiend*, 183; Dupouy, "Quelques réflexions sur le morphino-manie," 457.

11. Emmanuelle Retaillaud-Bajac, *Les paradis perdus: drogues et usagers de drogues dans la France de l'entre-deux-guerres* (Rennes, France: Presses Universitaires de Rennes, 2009), 298–299; Hehir, *Opium*, 637; Crowley, *Diary of a Drug Fiend*, 182; Kerr, "Alcoholism and Drug Habits," 83–84; Sollier's comment in discussion following Dupouy's presentation in "Quelques réflex-ions sur le morphinomanie," 465; "Further Note by Dr. Hogg on Cases of Addiction Treated," no date, PRO, MH 58/277.

12. Page, "Le traitement rationnel," 536; Dupouy, "Quelques réflexions sur le morphino-manie," 462–463; Jennings quoted in Scott, *Morphine Habit*, 17–18, 65.

13. Adams, "Report upon the Results"; Departmental Committee, *Report*, 15; "Precis of Evidence by F.S.D. Hogg, Resident Medical Superintendent, Dalrymplye House Retreat," no date, PRO, MH 58/277; Scott, *Morphine Habit*, 65; Miller, "Treatment of Morphinomania," 1596; Francis Hare, "The Withdrawal of Narcotics from Habitués," *British Journal of Inebriety* 8, no. 2 (1910), 87–88; Kerr, "Alcoholism and Drug Habits," 84; Jouet, *Étude sur le morphinisme*, 62; Page, "Le traitement rationnel," 539–540.

14. Page, "Le traitement rationnel," 537–538; Ghellerter, *Les toxicomanies*, 18; Crowley, *Diary of a Drug Fiend*, 183; Scott, *Morphine Habit*, 23; Departmental Committee, *Report*, 14–15.

15. Departmental Committee, *Report*, 13, 16; Adams, "Report upon the Results"; Scott, *Mor-phine Habit*, 16; Donald, "Treatment of Morphinomania," 411; Levinstein, *Morbid Craving*, 122; Edouard Kilidjian, *Prophylaxie médico-sociale des toxicomanies* (Paris: Jouve & Cie, 1935), 57; Page, "Le traitement rationnel," 537–538; Berridge, *Opium and the People*, 162; Kerr, "Alcoholism and Drug Habits," 58, 86–87.

16. "Sur le concours pour le Prix Deportes en 1891," *Bulletin de L'Académie de Médecine*, 3rd ser., 26 (1891), 586; Departmental Committee, *Report*, 17; "Further Note by Dr. Hogg on Cases of Addiction Treated," no date, PRO, MH 58/277; Ghelerter, *Les toxicomanies*, 68, 86.

17. Crowley, *Diary of a Drug Fiend*, 183–184; Cocteau, *Diary of an Addict*, 11.

18. Joseph Melling and Bill Forsythe, *The Politics of Madness: The State, Insanity, and Society in England, 1845–1914* (London: Routledge, 2006), 13–14; Kathleen Jobes, *Mental Health and Social Policy, 1845–1959* (London: Routledge & Kegan Paul, 1960), 15–40, 61–72; David Wright, *Mental Disability in Victorian England: The Earlswood Asylum, 1847–1901* (Oxford: Clarendon Press, 2001), 16–18; R. W. Lee, "Comparative Legislation as to Habitual Drunkards," *Journal of the Society of Comparative Legislation* 3, no. 2 (1901), 244–245; G. Hunt, J. Mellor, and J. Turner, "Wretched, Hatless, and Miserably Clad: Women and the Inebriate Reformatories from 1900–1913," *British Journal of Sociology* 40, no. 2 (June 1989), 245.

19. Virginia Berridge, "The Origins and Early Years of the Society, 1884–1899," *British Jour-nal of Addiction* 85 (1990), 993, 1000–1001; Willcox, "Norman Kerr Memorial Lecture," 1016; "Precis of Evidence by F.S.D. Hogg," PRO, MH 58/277; Berridge, *Opium and the People*, 165, 169–170.

20. Berridge, *Opium and the People*, 166–168; Andrew Rutherford, "Boundaries of English Penal Policy," *Oxford Journal of Legal Studies* 8, no. 1 (spring 1988), 135; Hunt, Mellor, and Turner,

"Wretched, Hatless," 245; Scott, *Morphine Habit*, 54; Willcox, "Norman Kerr Memorial Lecture," 1018; Robert Armstrong-Jones, "Drug Addiction in Relation to Mental Disorder," *British Journal of Inebriety* 12, no. 3 (Jan. 1915), 127, 139.

21. Home Office, "Brief Summary of the Contents of Certain American and Canadian Pamphlets Forwarded by the Home Office, Appendix I," PRO, MH 58/277; Willcox, "Norman Kerr Memorial Lecture," 1016; "Letter from Prison Commission to R. H. Crooke, Ministry of Health," November 6, 1924, PRO, MH 58/277.

22. One such "charitable institution" was the National Institution for Persons Requiring Care and Control, located in London, which is mentioned in "Letters, Notes, and Answers," *British Medical Journal*, April 16, 1910, 972. Another was run by the Church of England Temperance Society in the 1920s, as described in "Medical News," *British Medical Journal*, July 18, 1925, 152. "Committee on Drug Addiction: Second Meeting," October 31, 1924, PRO, MH 58/276; Adams, "Report upon the Results"; Walter Asten, "The Institutional Treatment of the Alcoholic Inebriate and the Drug Addict," *British Journal of Inebriety* 22, no. 2 (Oct. 1924), 48–49; Scott, *Morphine Habit*, 67; "Letter to Mr. Brock," August 6, 1924, PRO, MH 58/275.

23. W. E. Dixon's letter to the *Times* quoted in Berridge, *Opium and the People*, 271; "Addiction to Drugs," *British Medical Journal*, November 19, 1921, 856.

24. A New York Specialist [Aleister Crowley], "The Great Drug Delusion," *English Review* 34 (1922), 575–576.

25. T. Henderson to Home Office, November 20, 1922, reprinted in Home Office to Minister of Health, January 19, 1923, PRO, MH 58/275; Home Office to Minister of Health, January 19, 1923, PRO, MH 58/275; Berridge, *Opium and the People*, 270; Mold, *Heroin*, 14–15.

26. Henderson to Home Office, November 20, 1922, PRO, MH 58/275.

27. Home Office to Minister of Health, January 19, 1923, PRO, MH 58/275.

28. Departmental Committee, *Report*, 6–7; Home Office, "Brief Summary, Appendix I," PRO, MH 58/277.

29. Home Office, "Brief Summary, Appendix I," PRO, MH 58/277; Home Office to Minister of Health, January 19, 1923, PRO, MH 58/275.

30. Home Office to Minister of Health, January 19, 1923, PRO, MH 58/275.

31. "Note to Mr. Cleary," February 17, 1923, PRO, MH 58/275.

32. Ibid.; "Letter to Mr. Brock," August 6 1924; "Letter to Dr. Whitaker," March 13, 1923; all in PRO, MH 58/275.

33. "Letter to Dr. Whitaker," March 13, 1923, PRO, MH 58/275.

34. Berridge, *Opium and the People*, 272; "Note to the Secretary of the Treasury," September 24, 1924, PRO, MH 58/275; "Press Notice," no date, PRO, MH 58/275; "Statements in the Home Office Memorandum of Difficulties Experienced in Administration Which the Committee Are—Directly and Indirectly—Asked to Suggest a Means of Obviating or Diminishing," no date, PRO, MH 58/278; Departmental Committee, *Report*, 2, 20.

35. Departmental Committee, *Report*, 6–7, 10, 17–18, 20.

36. Ibid., 7, 18–19.

37. Ibid., 20–21; Mold, *Heroin*, 20.

38. Berridge, *Opium and the People*, 277; Terry M. Parssinen, *Secret Passions, Secret Remedies: Narcotic Drugs in British Society, 1820–1930* (Philadelphia: Institute for the Study of Human Issues, 1983), 194; Alfred R. Lindesmith, "The British System of Narcotics Control," *Law and*

Contemporary Problems 22, no. 1 (1957), 140; Bing Spear, "The Early Years of the 'British System' in Practice," in John Strang and Michael Gossop (eds.), *Heroin Addiction and Drug Policy: The British System* (Oxford: Oxford University Press, 1994), 12–13; Scott, *Morphine Habit*, 4.

39. John Strang and Michael Gossop, "The 'British System': Visionary Anticipation or Masterly Inactivity?" in Strang and Gossop, *Heroin Addiction and Drug Policy*, 342; Mold, *Heroin*, 10; Virginia Berridge, "The 'British System' and Its History: Myth and Reality," in John Strang and Michael Gossop (eds.), *Heroin Addiction and the British System, Volume 1: Origins and Evolution* (London: Routledge, 2005), 15.

40. Patricia Prestwich, *Drink and the Politics of Social Reform: Antialcoholism in France Since 1870* (Palo Alto, CA: Society for the Promotion of Science and Scholarship, 1988), 57–60; Lee, "Comparative Legislation," 249; "De la séquestration des alcooliques," *Annales Médico-psychologiques*, 5th ser., 7 (1872), 433; Pichon, *Le morphinisme*, 245; Gaudry, *Du morphinisme chronique*, 20–21.

41. Gaudry, *Du morphinisme chronique*, 68; "Chronique," *Annales Médico-psychologiques*, 6th ser., 10 (1883), 179; Dr. Blanche, "Procès de la femme Fiquet, accusée d'assassinat," *Annales Médico-psychologiques*, 6th ser., 10 (1883), 234–253; Garnier, "De l'état mental," 353–356; "Crimes et délits commis par les morphinomanes," *Annales Médico-psychologiques*, 7th ser., 18 (1893), 479; F. Senlecq, "Un cas de morphinomanie," *Annales Médico-psychologiques*, 8th ser., 1 (1895), 35; Pichon, *Le morphinisme*, 245; Société médico-psychologique, "Séance du 26 novembre 1888," 151–152; Ghelerter, *Les toxicomanies*, 24; Jouet, *Étude sur le morphinisme*, 16, 63–64; "Legal Measures for Facilitating the Treatment of Morphiomania," *Lancet*, May 20, 1899, 1397.

42. *Lancet*, "Legal Measures for Facilitating Treatment," 1397; Senlecq, "Un cas de morphinomanie," 34; Jean Colly, in *Journal Officiel de la République Française*, May 14, 1913, 566; Retaillaud-Bajac, *Les paradis perdus*, 241–255.

43. Retaillaud-Bajac, *Les paradis perdus*, 143–158; "Vingt-sept ans de morphinomanie, guérison 'spontanée' définitive," *Annales Médico-psychologiques*, 14th ser., 2 (1932), 209; M. Briand, "Démorphinisation par la 'suppression brusque,'" *Annales Médico-psychologiques*, 10th ser., 4 (1913), 584–585; Goudot, *Quelques considérations*, 11; Gheleter, *Les toxicomanies*, 25, 66, 87.

44. Retaillaud-Bajac, *Les paradis perdus*, 294–297, 300; Ghelerter, *Les toxicomanies*, 17–18, 36, 74; "Experiences of a Medical Morphinist," in Scott, *Morphine Habit*, 87–89; Gregory M. Thomas, "Open Psychiatric Services in Interwar France," *History of Psychiatry* 15, no. 2 (2004), 134–136.

45. Retaillaud-Bajac, *Les paradis perdus*, 301–303, Dupouy quoted on p. 302; Maurice Delaville and Roger Dupouy, "Procède de désintoxication rapide des morphinomanes par les émulsions de lipids," *Annales Médico-psychologiques*, 15th ser. (1935), 879; J. F. Buvat and H. Sauguet, "Un couple d' héroïnomanes, innocuité des doses massives d'héroïne," *Annales Médico-psychologiques*, 15th ser. (1939), 461.

46. Gheleter, *Les toxicomanies*, 35, 66; Retaillaud-Bajac, *Les paradis perdus*, 291–294, 298.

47. Antonin Artaud (trans. Helen Weaver), "Letter to the Legislator of the Law on Narcotics" (1925), in Susan Sontag (ed.), *Antonin Artaud, Selected Writings* (New York: Farrar, Straus and Giroux, 1976), 68–71; Antonin Artaud, "Sûreté générale: la liquidation de l'opium," *La Révolution Surréaliste* 1, no. 2 (Jan. 15, 1925), 20–21, emphasis in original.

48. Artaud, "Sûreté générale," 20–21, emphases in original.

49. Retaillaud-Bajac, *Les paradis perdus*, 77–94, 309–311; Ghelerter, *Les toxicomanies*, 131–132; Kilidjian, *Prophylaxie médico-sociale*, 51–54; I. Bussel, *L'état mental des toxicomanes* (Paris: Jouve & Cie, 1936), 78.

50. J. M. Dupain, "Chronique: à propos du procès des toxicomanes," *Annales Médico-psychologiques*, 12th ser., 1 (1923), 389–391.

51. Ghelerter and *Paris Soir* quoted in Retaillaud-Bajac, *Les paradis perdus*, 316, 318; Georges Boussange, *Le péril toxique en France* (Paris: Jouve & Cie, 1922), 42, 45; Ghelerter, *Les toxicomanies*, 165.

52. Ghelerter, *Les toxicomanies*, 24–25, 35, 43, 68–69; Dupouy, "Quelques réflexions sur le morphinomanie," 465.

53. Retaillaud-Bajac, *Les paradis perdus*, 315–316.

54. Ibid., 85, 315; *Gazette du Palais* (1932, 2nd Semester), 728.

55. Retaillaud-Bajac, *Les paradis perdus*, 311; René Leriche (trans. Archibald Young), *The Surgery of Pain* (1937) (Baltimore: Williams & Wilkins Co., 1939), 220–221, 460–461, 464; G. L. Tourade, "La santé publique et la lutte contre les stupéfiants," *Le Mouvement Sanitaire*, February 1934, 93.

56. Bussel, *L'état mental des toxicomanes*, 78; Kilidjian, *Prophylaxie médico-sociale*, 52; Dupain, "Chronique," 391; Retaillaud-Bajac, *Les paradis perdus*, 311.

57. Amélie Buvat-Cottin, "Considérations cliniques et thérapeutiques sur les toxicomanies," *Annales Médico-psychologiques*, 15th ser. (1936), 509; Ghelerter, *Les toxicomanies*, 87, 182, 186; Boussange, *Le péril toxique*, 74–76; B. J. Logre, *Toxicomanies* (Paris: Stock, 1924), 65.

58. Retaillaud-Bajac, *Les paradis perdus*, 319; Marcel Briand and Dr. Cazeneuve, "Les stupéfiants et la santé publique: la loi du 12 juillet 1916, doit-elle être reformée?" *Bulletin de la Ligue de l'Hygiène Mentale* 1, no. 1 (1921), 19; Logre, *Toxicomanies*, 72.

59. Kilidjian, *Prophylaxie médico-sociale*, 65; Ghelerter, *Les toxicomanes*, 170; Igor Charras, "L'état et les 'stupéfiants'—archéologie d'une politique publique répressive," *Les Cahiers de la Sécurité Intérieure*, 1998, 15–16.

60. Total population statistics are from B. R. Mitchell, *International Historical Statistics, Europe 1750–2000* (Houndmills, UK: Palgrave Macmillan 2003), 8. The prevalence estimate is based on the assumption of seven hundred addicts (the number estimated by the Home Office in 1935, when it first started keeping statistics) in a British population of approximately forty-five million. Using Retaillaud-Bajac's estimates for France, I assume five thousand addicts in a population of approximately forty-one million. Mold, *Heroin*, 20; H. B. Spear, "The Growth of Heroin Addiction in the United Kingdom," *British Journal of Addiction to Alcohol and Other Drugs* 64, no. 2 (1969), 247; Retaillaud-Bajac, *Les paradis perdus*, 229.

Indications are that although cocaine was more prevalent in France than in Britain, opiates were still the most prominent drugs. In Retaillaud-Bajac's sample of drug cases in Paris during the interwar period, 17.56 percent involved cocaine and less than 1 percent involved hashish, compared with 9.6 percent involving opium, 19.67 percent morphine, and 29.04 percent heroin. More than three-quarters of the cases documented in medical writings on addiction treatment in the interwar period analyzed by Retaillaud-Bajac involved addiction to opium or opiates. Retaillaud-Bajac, *Les paradis perdus*, 422, 424.

61. Spear, "Growth of Heroin Addiction," 248; Mold, *Heroin*, 20; Retaillaud-Bajac, *Les paradis perdus*, 65, 100–103, 112, 117, 126–134, 189, 193–194, 205–208, 376–378.

62. On institutions constructed to imprison, treat, and study drug addicts in the United States, see David F. Musto, *The American Disease: Origins of Narcotics Control* (New York: Oxford University Press, 1987), 204–206; Caroline Jean Acker, *Creating the American Junkie: Addiction Research in the Classic Era of Narcotic Control* (Baltimore: Johns Hopkins University Press, 2002).

Epilogue • Changes and Continuities

1. Carol Smart, "Social Policy and Drug Addiction: A Critical Study of Policy Development," *British Journal of Addiction* 79, no. 1 (1984), 36; Arnold S. Trebach, *The Heroin Solution* (New Haven, CT: Yale University Press, 1982), 108–111, 178; Duff Gillespie, M. M. Glatt, Donald R. Hills, and David J. Pittman, "Drug Dependence and Drug Abuse in England," *British Journal of Addiction to Alcohol and Other Drugs* 62, no. 1–2 (1967), 157–161; Robert Power, "Drug Trends since 1968," in John Strang and Michael Gossop (eds.), *Heroin Addiction and Drug Policy: The British System* (Oxford: Oxford University Press, 1994), 29; Bing Spear, "The Early Years of the 'British System' in Practice," in Strang and Gossop, *Heroin Addiction*, 8–12, 17–18; H. B. Spear, "The Growth of Heroin Addiction in the United Kingdom," *British Journal of Addiction to Alcohol and Other Drugs* 64, no. 2 (1969), 250–254; Ministry of Health and Scottish Home and Health Department, *Drug Addiction: The Second Report of the Interdepartmental Committee* (London: Her Majesty's Stationery Office, 1965), 5–6; David J. Pittman, "The Rush to Combine: Sociological Dissimilarities of Alcoholism and Drug Abuse," *British Journal of Addiction to Alcohol and Other Drugs* 62, no. 3–4 (1967), 341–342; John Witton, Francis Keaney, and John Strang, "They Do Things Differently over There: Doctors, Drugs, and the 'British System' of Treating Opiate Addiction," *Journal of Drug Issues* 35, no. 4 (2005), 781–782.

2. On the creation of new mechanisms for prescribing opiates to addicts, see Ministry of Health, *Drug Addiction*, 8–9; Alex Mold, *Heroin: The Treatment of Addiction in Twentieth-Century Britain* (DeKalb, IL: Northern Illinois University Press, 2008), 38; Philip Connell and John Strang, "The Creation of the Clinics: Clinical Demand and the Formation of Policy," in Strang and Gossop, *Heroin Addiction*, 167–177.

For more detail on prescribing practices in the 1970s, see Mold, *Heroin*, 43, 54–55; Power, "Drug Trends since 1968," 29–30; Gerry V. Stimson, "British Drug Policies in the 1980s: A Preliminary Analysis and Suggestions for Research," *British Journal of Addiction* 82, no. 5 (1987), 479; Michael Gossop and Marcus Grant, "A Six Country Survey of the Content and Structure of Heroin Treatment Programmes Using Methadone," *British Journal of Addiction* 86, no. 9 (1991), 1158; Susanne MacGregor and Lynne Smith, "The English Drug Treatment System: Experimentation or Pragmatism?" in Harald Klingemann and Geoffrey Hunt (eds.), *Drug Treatment Systems in an International Perspective: Drugs, Demons, and Delinquents* (Thousand Oaks, CA: Sage Publications, 1998), 71–72; John Strang, Susan Ruben, Michael Farrell, and Michael Gossop, "Prescribing Heroin and Other Injectable Drugs," in Strang and Gossop, *Heroin Addiction*, 196–199; Martin Mitcheson, "Drug Clinics in the 1970s," in Strang and Gossop, *Heroin Addiction*, 178–191.

3. Comment by Senator Pierre Marcilhacy, quoted in Alain Ehrenberg, "Comment vivre avec les drogues? Questions de recherche et enjeux politiques," *Communications* 62 (1996), 6–9; Tim Boekhout van Solinge, *Dealing with Drugs in Europe: An Investigation of European Drug*

222 Notes to Pages 177–180

Control Experiences: France, the Netherlands, and Sweden (The Hague: Boom Juridische uitgevers, 2004), 68, 70, 81–83, 190; Alain Ehrenberg, *L'individu incertain* (Paris: Calmann-Levy,
1995), 70–71; Henri Bergeron, *L'état et la toxicomanie: histoire d'une singularité française* (Paris:
Presses Universitaires de France, 1999), 23–26; Philippe Chossegros, "Prise en charge de la toxicomanie en France (une histoire)," *Gastroentérologie Clinique et Biologique* 31, no. 8–9 (2007),
4S44–4S45; Yann Bisou, "France: Drug Use and Supply Illegal, Possession Undefined—Situation Unsatisfactory?" in Nicholas Dorn and Alison Jameison (eds.), *European Drug Laws: The
Room for Manoeuvre* (London: DrugScope, 2001), 93.

4. The "therapeutic injunction" was first conceived in 1953, but procedures for implementing it were not established until the 1970 law. Laurence Simmat-Durand and Thomas Rouault,
"Injonction thérapeutique et autres obligations de soins," www.hopital-marmottan.fr/spip/
IMG/pdf/dossier-injonc.pdf (accessed January 2011). By the mid-1990s, the proportion of individuals eligible to receive the therapeutic injunction that actually received it rose to about six-
teen percent. Laurence Simmat-Durand, "Latest Trends in French Policy on Drugs," *European
Journal on Criminal Policy and Research* 6 (1998), 418, 425–426.

Boekhout van Solinge, *Dealing with Drugs in Europe*, 83–87; Ehrenberg, "Comment vivre,"
10–11; Henri Bergeron and Pierre Kopp, "Policy Paradigms, Ideas, and Interests: The Case of the
French Public Health Policy toward Drug Abuse," *Annals of the American Academy of Political
and Social Science* 582, no. 1 (2002), 39, 43; France Lert, "Drug Use, AIDS and Social Exclusion in
France," in Jean-Paul Moatti, Yves Souteyrand, Annick Prieur, Theo Sandfort, and Peter Aggle-
ton (eds.), *AIDS in Europe: New Challenges for the Social Sciences* (London: Routledge, 2000),
190–191.

5. Boekhout van Solinge, *Dealing with Drugs in Europe*, 68, 86–87, 190; Bergeron and Kopp,
"Policy Paradigms, Ideas, and Interests," 39, 40, 42; Gossop and Grant, "Six Country Survey,"
1156; Peter Baldwin, *Disease and Democracy: The Industrialized World Faces AIDS* (Berkeley:
University of California Press, 2005), 148–149; Bergeron, *L'état et la toxicomanie*, 43–138, detailed
discussion of the Pelletier report on pp. 90–93; "The 8th Conference on the Reduction of Drug
Related Harm, Paris, March 1997: A Brief Report," *International Journal of Drug Policy* 9, no. 1
(1998), 5; Ehrenberg, *L'individu incertain*, 70.

6. In 1990 there were about eighty thousand individuals who were dependent on opiates in
France, compared with about seventy-five thousand in Britain; yet in France, fewer than a hun-
dred patients received methadone, compared with between two and five thousand in Britain.
Gossop and Grant, "Six Country Survey," 1153–1157.

7. Ibid., 1157; "Report of the Conference of Ministers of Health on Narcotic and Psychotropic Drug Misuse, London, 18–20 March 1986," *British Journal of Addiction* 81, no. 6 (1986),
831–838; Stimson, "British Drug Policies in the 1980s," 486; Baldwin, *Disease and Democracy*,
142–152; Bruce Bullington, "Drug Policy Reform and Its Detractors: The United States as the
Elephant in the Closet," *Journal of Drug Issues* 34, no. 3 (2004), 698; Boekhout van Solinge, *Dealing with Drugs in Europe*, 24; Roy Robertson, "The Arrival of HIV," in Strang and Gossop, *Heroin
Addiction*, 91–92; Gerry V. Stimson, "AIDS and HIV: The Challenge for British Drug Services,"
British Journal of Addiction 85, no. 3 (1990), 329–339.

8. Virginia Berridge, "AIDS, Drugs and History," *British Journal of Addiction* 87, no. 3 (1992),
366–367; Robertson, "Arrival of HIV," 96–97; Mold, *Heroin*, 128–142; Witton, Keaney, and Strang,
"They Do Things Differently over There," 786–788; Edwards, "What Drives British Drug Poli-

cies?" 223; Gossop and Grant, "Six Country Survey," 1158; Jon E. Zibbell, "Can the Lunatics Actually Take over the Asylum? Reconfiguring Subjectivity and Neo-liberal Governance in Contemporary British Drug Treatment Policy," *International Journal of Drug Policy* 15, no. 1 (2004), 57–58; Steve Cranfield, Charlotte Feinmann, Ewan Ferlie, and Cathy Walter, "HIV and Drugs Services—The Challenge of Change," in Strang and Gossop, *Heroin Addiction*, 323–327; Gerry V. Stimson, "Harm Reduction in Action: Putting Theory into Practice," *International Journal of Drug Policy* 9, no. 6 (1998), 401–409; Gerry V. Stimson, "Minimizing Harm from Drug Use," in Strang and Gossop, *Heroin Addiction*, 253–255.

9. Bergeron and Kopp, "Policy Paradigms, Ideas, and Interests," 39, 44; Simmat-Durand, "Latest Trends in French Policy," 420; Philippe Mossé, "A System at Its Starting Blocks: Drug Treatment in France," in Klingemann and Hunt, *Drug Treatment Systems*, 202; Boekhout van Solinge, *Dealing with Drugs in Europe*, 91.

10. Henrion Commission's report quoted in Boekhout van Solinge, *Dealing with Drugs in Europe*, 70; Ministry of Justice guidelines discussed in Boekhout van Solinge, *Dealing with Drugs in Europe*, 92; X. Thirion, J. Micallef, K. Barrau, S. Djezzar, H. Lambert, J. L. Sanmarco, and G. Lagier, "Recent Evolution in Opiate Dependence in France during Generalization of Maintenance Treatments," *Drug and Alcohol Dependence* 61, no. 3 (2001), 281–285; Chossegros, "Prise en charge," 4S44; Maria-Patrizia Carrieri, D. Rey, A. Loundou, G. Lepeu, A. Sobel, Y. Obaida, and the MANIF–2000 Study Group, "Evaluation of Buprenorphine Maintenance Treatment in a French Cohort of HIV-Infected Drug Users," *Drug and Alcohol Dependence* 72, no. 1 (2003), 13–14; Simmat-Durand, "Latest Trends in French Policy," 421; Julien Emmanuelli and Jean-Claude Desenclos, "Harm Reduction Interventions, Behaviours and Associated Health Outcomes in France, 1996–2003," *Addiction* 100, no. 11 (2005), 1690–1700; Baldwin, *Disease and Democracy*, 149–150; Bergeron, *L'état et la toxicomanie*, 197–300.

11. Mold, *Heroin*, 156; Witton, Keaney, and Strang, "They Do Things Differently over There," 788–789; Michael Gossop and Francis Keaney, "Research Note—Prescribing Diamorphine for Medical Conditions: A Very British Practice," *Journal of Drug Issues* 34, no. 2 (2004), 441–449; John Henley, "French Seize Roddick Products; Body Shop Founder Denies New Hemp Range Promotes Drug Use," *Guardian* (Manchester), August 28, 1998; Boekhout van Solinge, *Dealing with Drugs in Europe*, 4–5, 68, 203–206; Paul H. H. M. Lemmens and Henk F. L. Garretsen, "Unstable Pragmatism: Dutch Drug Policy under National and International Pressure," *Addiction* 93, no. 2 (1998), 157–158; Tim Boekhout van Solinge, "Dutch Drug Policy in a European Context," *Journal of Drug Issues* 29, no. 3 (1999), 518; Hans C. Ossebaard and Govert F. van de Wijngaart, "Purple Haze: The Remaking of Dutch Drug Policy," *International Journal of Drug Policy* 9, no. 4 (1998), 267.

12. Michael Rowe, Patricia Benedict, Dave Sells, Tom Dinzeo, Charles Garvin, Lesley Schwab, Madelon Baranoski, Vincent Girard, and Chyrell Bellamy, "Citizenship, Community, and Recovery: A Group- and Peer-Based Intervention for Persons with Co-Occurring Disorders and Criminal Justice Histories," *Journal of Groups in Addiction and Recovery* 4, no. 4 (2009), 224–244; Betty Ford Institute Consensus Panel, "What Is Recovery? A Working Definition from the Betty Ford Institute," *Journal of Substance Abuse Treatment* 33, no. 3 (2007), 221–228.

13. Betty Ford Institute Consensus Panel, "What Is Recovery?" 222, 225; Thomas McClellan, "What Is Recovery? Revisiting the Betty Ford Institute Consensus Panel Definition," *Journal of Substance Abuse Treatment* 38, no. 2 (2010), 200–201; William L. White, Pat Taylor, and Carol

McDaid, "Recovery and Citizenship," 2010, www.williamwhitepapers.com (accessed February 2011); White quoted in Maia Szalavitz, "The Addiction Files: How Do We Define Recovery?" *Time Healthland,* September 20, 2010, http://healthland.time.com (accessed February 2011).

14. United Kingdom Drug Policy Commission, *UK Drug Policy Commission Recovery Consensus Group: A Vision of Recovery* (London: UK Drug Policy Commission, 2008).

15. White, Taylor, and McDaid, "Recovery and Citizenship."